The
Secret
Lives
of the
Elements

For Mark Whiting

First published in Great Britain in 2021 by
Greenfinch
An imprint of Quercus Editions Ltd
Carmelite House
50 Victoria Embankment
London EC4Y 0DZ

An Hachette UK company

Publisher: Kerry Enzor
Project editor: Anna Southgate
Design: Luke Bird
Illustrations: Jo Parry
Production: James Buswell

A CIP catalogue record for this book is available from the British Library

HB ISBN 978-1-52941-274-1
eBook ISBN 978-1-52941-275-8

10 9 8 7 6 5 4 3 2 1

Typeset in Bill Corporate Narrow and Adobe Garamond Pro
Printed and bound in China by C&C Offset Printing Co. Ltd.

Papers used by Greenfinch are from well-managed forests
and other responsible sources.

MIX
Paper from
responsible sources
FSC® C008047

The
Secret
Lives
of the
Elements

Kathryn
Harkup

greenfinch

Contents

Introduction

Just as illustrated alphabets feature on the walls of nurseries, posters of the periodic table adorn most science classrooms and are a familiar sight from our school days. While the illustrated alphabets of our infancy showed us the building blocks of language, the periodic table is a summary of the building blocks of life – the elements that make us and all the things around us.

Of the many ways to display these elements, one version is more familiar than most and takes shape as a long, low castle with a tower at each end: each element forms one of the many bricks in its walls. You may only have glanced at the table, but you would know it anywhere. Such a ubiquitous image easily merges into the background, and yet this stunning creation conveys an extraordinary amount of information in its one simple image and deserves closer scrutiny.

To me, the periodic table is a family photograph with all the branches of an extended family brought together for one gigantic wedding/funeral/baptism. Everyone is there – the conventional and dysfunctional family members, the extroverts and introverts, the standoffish and the friendly. Some elements get along with each other very well, a few are best kept apart, and a handful shun all but their own company. Families often have common physical traits and quirks of personality, and this extended family of elements is no exception. By grouping elements together by the characteristics that connect them, the periodic table manages to reflect the complex kinships between different branches of the family in one coherent design.

In broad terms, there are two types of element: metals, two-thirds of all the elements collected on the left and centre of the table; and non-metals, the remaining one-third occupying the top right. At a more detailed level of organization the elements are grouped by specific chemical traits. Close relatives, with similar characteristics, find themselves in columns with their own family name – the alkaline earth

metals, the chalcogens, the halides. Neighbouring columns are more like cousins, similar in some respects but having more in common with members of their own group. The further apart two elements are placed in the periodic table, the more different they are from each other.

The first column on the left contains the exuberant alkali metals, full of life and happiest in company. The last column, on the far right of the table, is made up of quiet, self-contained noble gases that want to be left alone. The middle of the table, the low, flat block that forms the body of the castle, is full of brash, colourful transition metals that are impossible to ignore. Below the main table, two rows of elements – the lanthanides and actinides – are distinct from the rest but remarkably similar to each other. These two tight-knit teams of siblings have caused all kinds of headaches and heartaches for the scientists who have studied them. What is so brilliant about this sophisticated, yet simple, system of organization is that, even for the least familiar elements, the merest glance at the table reveals clues about their appearance, alongside a wealth of information about how they will interact with others.

Beyond the familial similarities, each element is an individual with its own personality and characteristics. The intention of this book is not to present an overview of the periodic table but offers, instead, fifty-two elements with a story to tell. There are tales of the science that makes an element special and accounts of surprising discoveries. Some elements are relatively anonymous or new arrivals; others, already familiar, are seen in a new light; and old friends have secrets to share. Welcome to *The Secret Lives of the Elements*. Come and meet the family.

The Periodic Table

Group I	2	3	4	5	6	7	8
H							
Li	Be						
Na	Mg						
K	Ca	Sc	Ti	V	Cr	Mn	Fe	C
Rb	Sr	Y	Zr	Nb	Mo	Tc	Ru	R
Cs	Ba	Lu	Hf	Ta	W	Re	Os	
Fr	Ra	Lr	Rf	Db	Sg	Bh	Hs	N

Lanthanides	La	Ce	Pr	Nd	Pm	S
Actinides	Ac	Th	Pa	U	Np	F

11	12	13	14	15	16	17	18
							He
		B	C	N	O	F	Ne
		Al	Si	P	S	Cl	Ar
Cu	Zn	Ga	Ge	As	Se	Br	Kr
Ag	Cd	In	Sn	Sb	Te	I	Xe
Au	Hg	Tl	Pb	Bi	Po	At	Rn
Rg	Cn	Nh	Fl	Mc	Lv	Ts	Og

Gd	Tb	Dy	Ho	Er	Tm	Yb	Lu
Cm	Bk	Cf	Es	Fm	Md	No	Lr

Elem

The
ents

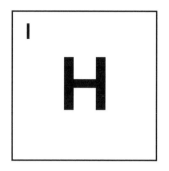

I

Hydrogen
The Misfit

Hydrogen is the original element. Born shortly after the Big Bang, it is one of a kind. Though it has similarities to elements in the first group and the seventh group of the periodic table, it does not really belong to either. It is distinct and special. Hydrogen may not be like other elements, but that is not hydrogen's fault, it was simply made that way.

Every atom of every element is made up of the same three things: protons, electrons and neutrons. Everything about an element is defined by the number and ratio of these three components. These basic building blocks determine both the identity and the behaviour of the atom. Hydrogen is hydrogen because it has one positive proton at its centre or nucleus, if it had any more it would be something else. To balance the positive proton, a negative electron circles around it. The result is a single atom that is neutral overall.

The third component is the neutron. Neutrons are there to hold all the positive protons together in the compact nucleus of the atom. But hydrogen, with only one proton, has no real need of neutrons and, for the most part, gets on perfectly well without them. This is just one of the many features that makes hydrogen that little bit different to all the other elements.

Hydrogen
Non-Metal

H

Melting Point
−259°C (−435°F)

Boiling Point
−253°C (−423°F)

Group   Period

I   I

If protons give an atom an identity, it is the electrons that give it character. Changing the number of protons in an atom is difficult and requires quite specific conditions for it to occur. Electrons, on the other hand, can be shared, donated, stolen and moved about easily. Hydrogen only has one electron to play around with, but it's what you do with it that counts.

When a hydrogen atom loses its electron, all that is left is a tiny, bare proton – H^+ – an unimaginably little thing with a positive charge. It may be small, but it is mighty. H^+ is what gives acids their bite. It is what makes dropping the toaster in a bathtub such a bad idea, because H^+ helps water conduct electricity. This single proton, unhindered by an electron, can attach to molecules to change their behaviour. In the presence of H^+, chemical reactions occur when before they could not. Molecules that will not dissolve in water can disappear into solution when an H^+ is around to help.

If losing its electron was what hydrogen always did, things would be a lot simpler, but it can also gain an electron to become H^-. And, though H^- has its uses, it is when hydrogen shares electrons that it really comes into its own. The hydrogen bond, a tenuous bridge of electrons shared between three atoms with hydrogen in the middle, is a game changer. Hydrogen bonds make life as we know it possible. They keep water liquid and ice afloat. These special bonds are strong enough to hold together strands of DNA, but weak enough for them to be pulled apart so that DNA can be read and copied.

The periodic table is organized based on similar properties, but hydrogen's versatile nature makes it difficult to place. A single negative charge is the defining characteristic of the elements in group 17. And some group-17 elements are gases, like hydrogen, but that is about where the similarities end. For this reason, hydrogen is sometimes, but not often, found at the top of group 17 on some periodic tables.

By contrast, having a single positive charge is the main feature of the first group in the periodic table, the alkali metals. Hydrogen is therefore also often seen hovering around the top of this group, even though it has few other characteristics in common with the group's other members. The most glaring difference is that hydrogen is a gas and not a metal like all the others in group 1. Many scientists think hydrogen atoms could become a metal if they could just squeeze them together tight enough, but no one here on Earth has yet achieved the required pressure to prove it. When placed with group 1, periodic table designers often give hydrogen a different colour, or leave a little gap, to highlight its uncomfortable position.

Hydrogen might be a difficult element to place but being different is what makes it so important. And when hydrogen makes up around three-quarters of all the atoms in the known universe, maybe it isn't hydrogen that is the odd one out, maybe it's all the others.

2

He

Helium
The Loner

Helium is the wallflower of the periodic table. Most elements interact with each other to form compounds, some more enthusiastically than others, but helium will have nothing to do with other elements (or even other atoms of itself). A few elements decay so soon after they are created that they simply do not exist long enough to undergo a chemical reaction, but helium has been around since the beginning. It condensed out of the broiling mass of matter and energy that was the Big Bang at around the 300,000-year mark. Even with time on their side, it is possible that some of the helium atoms that popped into being billions of years ago have never formed a chemical bond with another atom in their entire existence.

Helium is at the head of a group of elements known as the noble gases, a name that gives them an air of smug superiority they frankly do not deserve. Helium is not lazy, it does not look down on other elements as unworthy of interaction, it is simply that a helium atom is perfectly happy by itself.

The two protons at the centre of every atom define it as helium. Some 99.9998% of the helium atoms in your birthday balloon will also have two neutrons alongside those protons to form the nucleus.

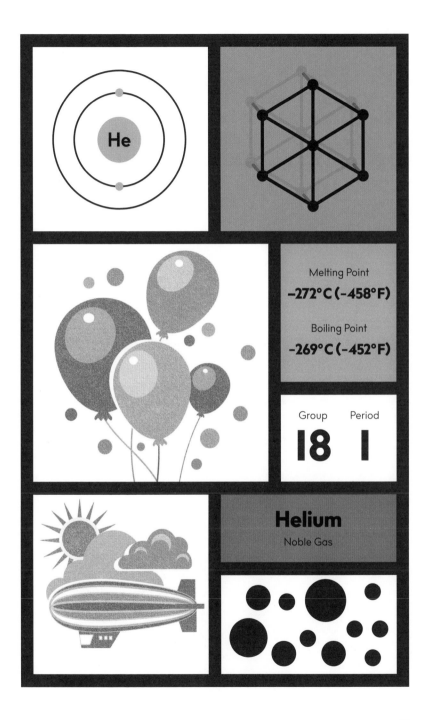

He

Melting Point
−272°C (−458°F)

Boiling Point
−269°C (−452°F)

Group   Period
18   **1**

Helium

Noble Gas

But it is the two electrons spinning around the protons and neutrons that make helium so content.

Electrons in all atoms sit in expanding shells around the central nucleus, and there are strict rules as to who sits where. Like teachers guiding students into the assembly hall, electrons fill up from the front, occupying the shells closest to the nucleus first, and there can be no gaps or overfilling before electrons start to occupy the next shell. A perfectly filled shell is both aesthetically pleasing and more energetically stable than a messy part-filled shell. Atoms that do not have perfectly filled shells donate, steal or share electrons from others in order to try and fill in the gaps or get rid of a few untidy extras. This is essentially what all chemical reactions are doing: trying to achieve the most stable arrangement of electrons through bonding with other atoms.

The first shell, being so close to the tiny nucleus, is also the smallest and only has room for two electrons. Helium, therefore, has a perfectly formed first shell. Adding an electron, or taking one away, would spoil the symmetry. It takes tremendous amounts of energy, such as that of high temperatures or high voltages, to budge these electrons out of their comfortable position – more energy than chemical reactions can provide. It is generally accepted that there simply are no compounds of helium. Its lack of interaction with other atoms means it rarely makes its presence felt. It is hardly surprising that helium, so unreactive and unnoticeable, was totally overlooked on Earth for so long. It was first spotted by chance, when someone stared at the sun.

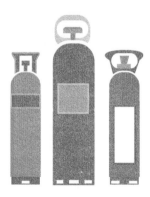

In 1868, the astronomer Pierre Jules Janssen schlepped halfway across the world to make observations of a solar eclipse. At the top of a signalling tower in Guntur, in eastern India, he set up a spectroscope, an instrument that can split the colours of light into fantastically detailed rainbows. The coloured bands of light obtained by spectroscopes are emitted by electrons that have been moved out of their regular shells by absorbing energy, and then fall back to their comfortable positions, releasing the energy they absorbed as light. The shells of each element are slightly different and so the bands of light they emit also vary – like a brightly coloured barcode that characterizes each element. Among the lines that Janssen recorded was one bright-yellow band that had not yet been assigned to any element, but he failed to take much notice of it.

A few months later another scientist, Norman Lockyer, travelled as far as his own back garden to make similar observations of the sun. He observed the same bright-yellow band that had no known element to answer for it. So, Lockyer declared there was a new element and named it helium, after the Greek *helios* for sun, thinking perhaps it did not occur on Earth. Most responses to this fantastic discovery, though confirmed by Janssen's earlier observations, were mocking. Whoever heard of an extraterrestrial element? But in 1895 helium was found to be very much Earth-bound, trapped in a sample of the mineral cleveite, completely unnoticed until William Ramsay found it with his spectroscope.

Lithium
The Tranquillizer

In 1929, Charles Leiper Grigg introduced a new health drink to the world. Despite the considerable hindrance of an unwieldy name (Bib-Label Lithiated Lemon-Lime Soda) and launching two weeks before the Wall Street Crash, bottles of Grigg's new beverage sold well. The name was soon shortened to 7Up Lithiated Lemon Soda and, by 1936, was just plain 7Up.

Theories about the drink's name abound. A few people thought it was because 7Up has a pH of 7, but it does not and never did. Other sources say it was due to the seven ingredients that went into Grigg's refreshing drink. But chemists prefer to believe it is a reference to just one of the original ingredients – lithium, with its mass number of 7.

Grigg's choice of lithium was deliberate. Tapping into the patent medicine market that was popular at the time, he originally promoted the drink as a hangover cure, though its efficacy in this regard is doubtful. The health benefits of lithium salts had been touted for decades. In the late nineteenth century, spa towns had grown up around wells that contained lithiated water. Anecdotes of the water curing everything from dementia to rheumatism sent people flocking to swallow it and bathe in it. Grigg was jumping on an already overloaded bandwagon.

Lithium
Alkali Metal

Melting Point
180°C (351°F)

Boiling Point
1342°C (2448°F)

Group
1

Period
2

Li

Tiny amounts of lithium are naturally present in our bodies simply because it is present in our environment. It is a group 1 metal, like sodium and potassium, and they all have similar chemistry. When scientists found that excess sodium contributed to hypertension and heart disease, doctors prescribed lithium salts in place of the sodium salt normally added to our food. Lithium was also found to dissolve crystals of uric acid that cause the excruciatingly painful attacks associated with gout, and thus more lithium prescriptions were written.

It *is* true that lithium will help dissolve uric acid crystals, though sadly the amount needed to dissolve the troublesome crystals in gout patients would be toxic. It is also true that lithium is similar to sodium and potassium, but it is not the same. Sodium and potassium carry out important functions in the body, but subtle differences in chemistry mean that lithium does not perform those same functions as effectively. In fact, lithium has no known biological role and even small amounts can easily disrupt and damage the body. By 1948 the dangers of too much lithium were well known and it was banned from all beverages, including 7Up. The following year the sale and medical use of lithium was also banned. The dangers of lithium medicine were very real, but the potential benefits had been overlooked.

In 1949 an Australian psychiatrist, John Cade, started experimenting with lithium urate. Uric acid was known to be psychoactive, but it did not dissolve very well in water. So, Cade used lithium to make lithium urate, which was soluble and could therefore be injected. The excitable guinea pigs that received the injections became calm. Cade had found that lithium could control mania and promoted its use as a tranquillizer. The rest of the medical world, wary of the potentially lethal consequences of even small overdoses of lithium, was slow to catch on.

Lithium is an unusual drug in many respects. The vast majority of medicines are chemical compounds of one sort or another, but for lithium drugs it is the element itself that does the work. In correct doses, and with careful monitoring, it can be an extremely effective way of treating bipolar disorder, severe depression and schizophrenic disorders. Other psychoactive drugs can create states of euphoria, but not lithium. It stabilizes mood but can also feel like being cut off from the world and, for some, the sense of detachment is too great a price to pay. For something that has no naturally occuring biological role, lithium has a remarkable effect on the body and mind.

Lithium helps to release chemicals serotonin and dopamine from nerve cells. It also changes the number of receptors for these chemicals in the brain. But exactly which of these effects is the one that reduces suicidal tendencies and relieves the misery of some psychiatric disorders is unknown. Grigg's original recipe perhaps helped relieve the angst of the stock market crash, but at the risk of poisoning his customers.

Beryllium
The Space Crusader

Beryllium is precious. In its pure form it is nothing much to look at, an unremarkable grey metal, but this unprepossessing appearance disguises a life of glamour and riches. Value is given to things for many reasons – beauty, utility, rarity – and beryllium has all three, they're just not always apparent. It is fitting that the first clues to beryllium's existence were found in a precious stone.

The eighteenth-century Enlightenment prompted insatiable curiosity about anything and everything. Some looked up at the stars in search of the secrets of the universe, some peered inside the human body to reveal the mystery of life, others examined rocks to track down the truth about nature. A few chemists studied precious jewels, perhaps hoping to find some wondrous substance that made these stones so alluring. Louis Nicolas Vauquelin, a French chemist, turned his attention to beryls, a class of gemstones that include aquamarines and emeralds, and are characterized by the wide variety of colours they display.

In 1798, Vauquelin pounded fine Peruvian emeralds to dust in search of their elusive alluring ingredients. What he found was an abundance of very ordinary silica, or sand, and common-or-garden alumina. Destroying these jewels robbed them of their financial worth, but it also

Beryllium

Alkaline Earth Metal

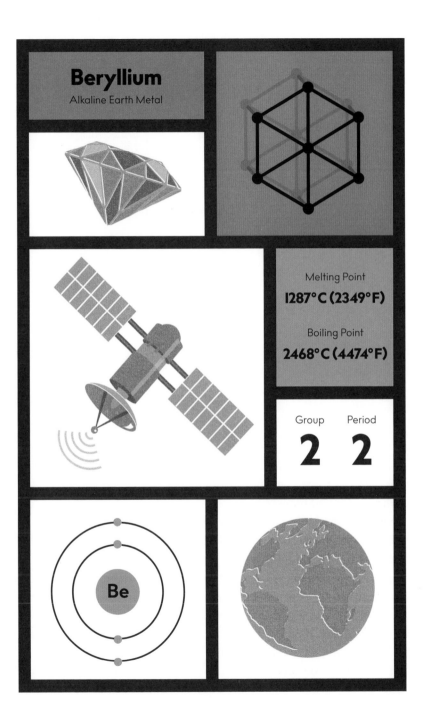

Melting Point
1287°C (2349°F)

Boiling Point
2468°C (4474°F)

Group Period

2 **2**

Be

revealed scientific riches. For there was something else lurking in the chemical remains of his destructive experiment. This something tasted like sugar, but it certainly was not sugar. Vauquelin named the mysterious substance glucina, for its sweet taste. He did not know what glucina was, but he was certain it contained a metal, which he proposed to call glucinum. The names did not catch on.

As more and more valuable gemstones were sacrificed to scientific curiosity, it became apparent that many contained the same basic ingredients. Vauquelin's sweet substance came to characterize the beryl class of minerals and precious stones, it was therefore named beryllia. The metal component of beryllia, when it was eventually isolated in 1828, was therefore named beryllium.

If scientists had hoped for a beautiful reflective metal that gave these jewels their sparkle, or a brilliantly coloured metal that gave beryls their rainbow hues, they were to be disappointed. Those first few dull, grey grains of beryllium that were produced in test tubes must have been underwhelming, to say the least. Furthermore, these flecks of metal could only be extracted in appreciable amounts from beryls and, though there are more than one hundred different types of beryl gemstone, most of them are rare. Obtaining the pure metal was a laborious process and destroying these beautiful crystals to extract their drab metal ingredient hardly seemed worth it. Its addition to the growing list of elements was a significant scientific achievement, but it needed a use.

Beryllium is brittle and difficult to work with, not least because beryllium dust is extremely toxic, meaning this metal has only specialist applications. But this does not relegate the element to obscure scientific niches. Beryllium's strength, resistance to heat and low density make it perfect for the aerospace industry. Boring beryllium, after decades languishing in obscurity, now leads an adventurous life in space, where it is used to give resilience to rockets, satellites and space telescopes.

Many elements have a high value placed on them shortly after their discovery. Often, as more applications are found and methods of extraction improve, prices drop, and an element's status can change from precious to mundane. Not so for beryllium. Its rarity means, even with the advances in technology, it is unlikely to be put to everyday use. This scarcity is due to the inherent properties of its atoms: beryllium may be strong on the outside, but it has a fragile heart.

Most elements are manufactured in stars, where the intense heat and gravitational pressure can force atomic nuclei together. As more and more protons are pushed into the tiny nuclei, new elements are created. But the nuclei of three elements do not survive for very long inside these vast stellar element factories: lithium, beryllium and boron. These three are quickly smashed into other nuclei to make other elements. It is believed that most of the beryllium in the universe was made well away from the intense environment of stars. Instead, it is thought to originate in interstellar dust clouds where cosmic rays smash apart heavier elements to leave beryllium behind in the atomic debris. While most elements are mass-produced, beryllium will always be a rarity produced by cosmic chance.

5

B

Boron
The Rocket Scientist (not)

Once upon a time there were two countries who were the bitterest of enemies. They did not want to fight each other but neither could they be reconciled. Instead, they waged a scientific war, a battle of technological superiority. Each country sent its agents to spy on the other, to check on each other's progress and discover any clues that might give them an advantage. One spy found his way to the edge of a military facility and hid himself away to watch the launch of the enemy's latest rocket. To his astonishment he saw green flames burst from the rocket's thrusters and rushed away to report his strange observation.

When the report was received back in his home country, the scientists scratched their heads and stared at the periodic table. Green flames? They knew that each element burned with a characteristic colour. Rockets had orange flames because they burned nitrogen, hydrogen and carbon compounds with oxygen. But green? And what shade of green? The spy's message contained no further clues.

There are quite a few elements that burn with a green flame. Could the green colour indicate that the enemy had made a new copper or barium alloy to build its thrusters? Or maybe it was not the rocket itself that was the source of the colour – what if it was the fuel?

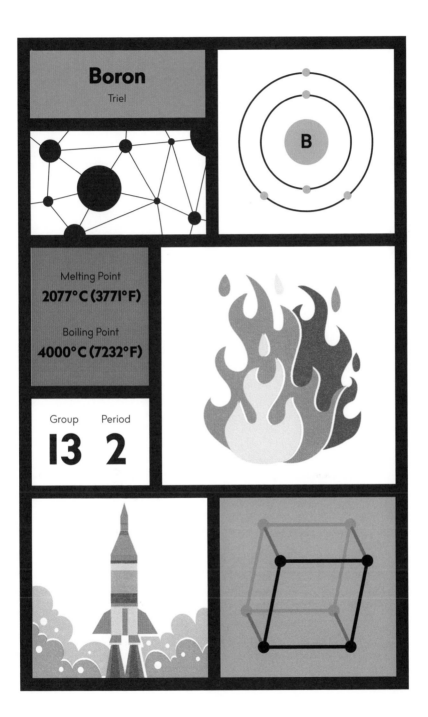

Boron

Triel

Melting Point
2077°C (3771°F)

Boiling Point
4000°C (7232°F)

Group
13

Period
2

Most conventional fuels are compounds called hydrocarbons, made up of carbon and hydrogen. These molecules react with oxygen to release a lot of energy and produce the yellow-orange flames most people are familiar with. But, if you look at the periodic table, right next to carbon is boron. Being neighbours, they have some chemical similarities, but also some differences. Like carbon, boron can form compounds with hydrogen, called boranes. They are similar to hydrocarbons in that they burn with oxygen, but they are much more reactive and release more energy for their weight. And, most importantly, boranes burn with a green flame.

Clearly the enemy had found a new rocket fuel that could give them that all important technological edge. Huge amounts of time and resources were ploughed into borane research to try and catch up and maybe even outpace the enemy. It was a steep learning curve.

Elements in the same group of the periodic table can be very similar. Neighbouring groups, however, are more like different branches of the same chemical family tree. Boron is carbon's weird cousin. Carbon is brilliantly balanced to share its electrons neatly and efficiently to form hydrocarbons. These molecules are stable enough to store, but reactive enough that only a small spark is needed to release their stored energy.

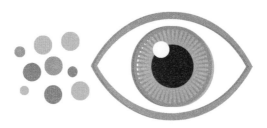

Boron is similar to carbon but with a few bits missing. Boron, even when it bonds with hydrogen atoms to form boranes, is a few electrons short of a full set, leaving it edgy and prone to grabbing more if the opportunity arises – one such opportunity being if it comes into close proximity to oxygen. Oxygen is rich in tantalisingly available electrons and boron does not hold back. Many boranes will catch fire as soon as they are exposed to the oxygen in the air. This is great when you want to launch a rocket, but nerve-racking for the staff who have to produce and store the fuel as well as fill up the rocket's tanks.

When hydrocarbons burn, they produce water and carbon dioxide. Admittedly carbon dioxide is no friend to the environment, but as far as engines and rockets are concerned, it is perfect because it is a non-toxic gas that disperses easily. The boron equivalents of hydrocarbons are toxic, and when they burn, they produce water and boron trioxide, a glassy solid that clogs engines and damages the blades of jet engines. To the rocket scientists, the huge gains in energy that boranes could offer seemed to outweigh the risks, so more time and money was thrown at the project. Workarounds were found and adaptations made, but still more issues were found. The scientists and engineers were exasperated. How could their enemy have made this technology work?

At the end of their tether the scientists demanded the spy be recalled so they could interview him. They begged him for any clues or details that could help them out of the never-ending labyrinth of technical hitches. It was during this detailed questioning that it emerged that the spy was colour blind. The flames he had seen all those years before had been orange.

Carbon
The Great Adaptor

Carbon is the Swiss Army knife of the periodic table, the Leonardo da Vinci of the elements, able to adapt and excel in a staggering number of roles and situations. At least one-third of any chemistry course will be devoted to the science of this one element and its compounds. To try and cover all of carbon's capabilities in one short chapter would be impossible. To try and outline just a few details of only two forms of pure carbon is foolhardy, but let's try it anyway.

As a pure element, carbon can take on several distinct guises. The atoms do not change, it is what you do with them. The two most common arrangements of carbon atoms result in graphite and diamond. Though they are made of exactly the same stuff, they are almost polar opposites. Graphite is the Clark Kent to diamond's Superman.

The properties of diamond are well known. Songs have been sung about them, they have starred in films and featured in novels. But we should not always take Dame Shirley or Ms Monroe at their word. Diamonds aren't always forever and these rocks *can* lose their shape.

Each diamond is an incompressible single crystal of carbon. Every carbon atom in a diamond is bonded to four other carbon atoms to form a huge, three-dimensional lattice. It is this compact and efficient

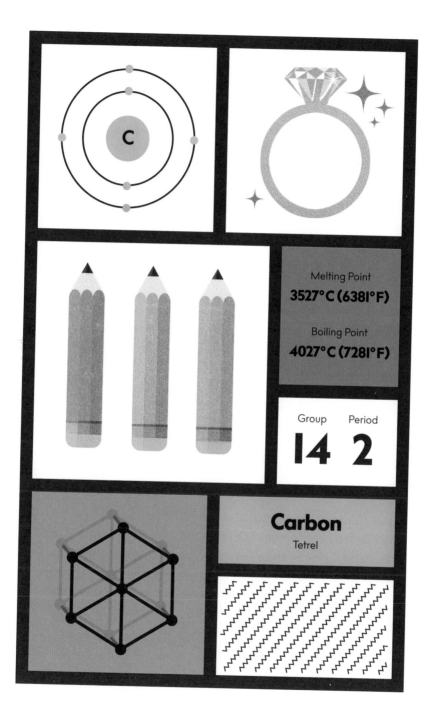

Melting Point
3527°C (6381°F)

Boiling Point
4027°C (7281°F)

Group Period
14 2

Carbon
Tetrel

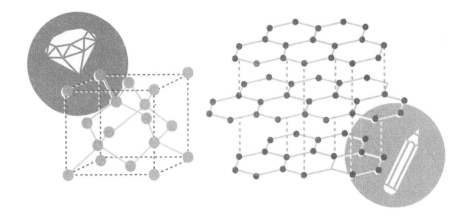

way of connecting atoms that gives diamonds their strength and beauty. But the precious stones that have come to symbolize eternity and stability are only metastable when compared to humble graphite.

The carbon atoms in graphite are each bonded to only three others to form honeycomb-like layers. Each carbon atom may form fewer bonds, but these bonds are stronger than those in diamond, meaning graphite is the most stable form of the element. These strongly bonded sheets of atoms are piled on top of each other, with only weak forces between the individual sheets, like a stack of paper. This arrangement gives graphite its very different, but no less impressive or desirable, properties.

Sometime before 1565, deposits of very pure plumbago were found in Borrowdale, Cumbria. Plumbago was the ancient name for graphite, from the Latin *plumbum* meaning lead, because it looks very similar to lead ores. Lumps of this mineral had been used to mark the local sheep, but people soon realized that it was far too precious a commodity to waste on livestock. A mine was established and the money rolled in.

The weak forces between graphite layers can easily be overcome with gentle friction. It was how the plumbago made its mark on the Borrowdale sheep, from layers of graphite wiped off on their fleece. It was also why, in the late sixteenth century, traders were selling Borrowdale plumbago to Michaelangelo's art school in Italy. And, it is why it was given the name graphite, after the Greek for 'writing stone'.

Graphite's slippery layers mean it can be shaped easily by tools, but that shape is remarkably resilient to heat. The strong bonds within graphite's layers stop it from melting, even when molten lead is poured on it. So, the English started using graphite to line the moulds for their cannonballs. The smoother, rounder ammunition travelled further and was more accurate, giving them a notable military advantage.

As the only source of plumbago in England, the Borrowdale mine became of national importance. Guards were posted and laws were passed to protect the valuable commodity buried in the Cumbrian hills. And while the value of plumbago was well appreciated, the identity of the material was more mysterious. Though the substance looked like a lead ore, it clearly was not.

Scientists were also stumped as to what diamonds could be made of, but their near indestructibility made it almost impossible to prize them apart to investigate. Certainly no one thought soft, grey-black graphite could be made of the same stuff as brilliant, hard diamond. That was until Antoine Lavoisier came along with a huge lens and a lot of scientific determination.

In 1772, Lavoisier concentrated the sun's rays through his lens and focused the beam on a diamond. It burned away to nothing, just as graphite will burn in intense heat. It was like Clarke Kent taking off his glasses. It still took another twenty-five years and more experiments, this time carried out by Smithson Tennant, to convince the Lois Lanes of the scientific world that graphite and diamond were the same thing but in very different guises.

Nitrogen
The Fertilizer

All life on Earth is dependent on the availability of nitrogen. Not that the element is in short supply, far from it: every breath we take is around 78 per cent nitrogen, but we breathe it out again unchanged and unused. The nitrogen atoms in the air that we breathe are tightly bound together in pairs, self-contained and chemically indifferent to everything around them. Instead, life needs nitrogen that has been combined with other elements, carbon, oxygen and hydrogen. Access to supplies of usable nitrogen can mean the difference between famine and untold riches, and in the mid-nineteenth century some were prepared to go to war over it.

Until the early twentieth century, when chemists devised industrial methods for prising apart nitrogen atoms in the air, all life was reliant on nature to do the work. Nature has two methods for breaking apart nitrogen pairs: brute force by lightning, or gentle persuasion by bacteria. Lightning is infrequent and haphazard, but it does convert nitrogen into nitrates that get soaked into the soil. Some specialized bacteria can convert nitrogen from the air into nitrates via ammonia, but these bacteria are only found associated with the roots of certain plants, such as clover and peas.

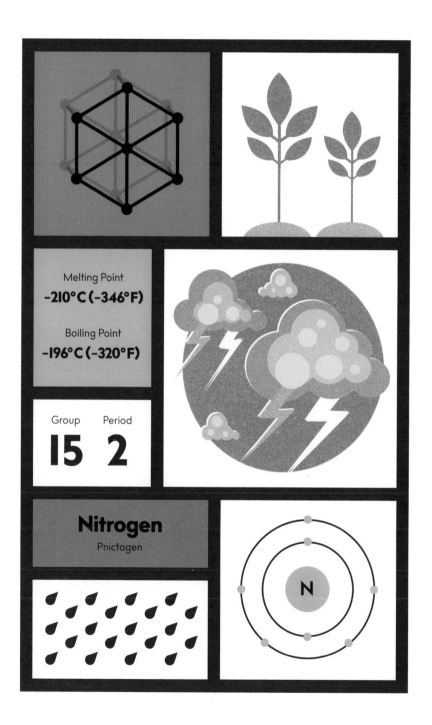

Melting Point
−210°C (−346°F)

Boiling Point
−196°C (−320°F)

Group
15

Period
2

Nitrogen
Pnictogen

N

Plants use the nitrates in soil to make all the nitrogen-containing compounds they need to survive and thrive. These nitrogen compounds are passed on up the food chain when animals eat plants. Life constantly depletes the amount of available nitrogen from the soil, but death returns it. When anything living dies, decay processes return nitrogen to the soil. An equilibrium is soon reached.

Agriculture disrupts this equilibrium. Repeated growing and harvesting of crops in the same soil can quickly exhaust the nitrate supplies. Waiting for a thunderstorm to replenish stocks is not practical. Growing plants with nitrogen-fixing bacteria in their roots may mean a field lies fallow for a year. Manure from farm animals can be used to replenish some of the lost nitrogen because animals eat far more nitrogen in their diet than they need, and the excess is excreted as waste. However, it is a cycle of ever-diminishing returns. Fertilizer is needed to maintain crop yields, but natural sources are difficult to find.

The Chinchas are three tiny islands off the coast of Peru, inhabited only by seabirds. They were once held in reverence by the native people of Peru. Only occasionally would they sail the six miles out to the islands to collect small amounts of the dusty white *huanu* that covered the granite rocks. Huanu was considered as valuable as gold because it made their crops of maize grow abundantly even in the poorest soils.

Colonizers were slow to realize the significance of the Chinchas. Explorers took samples of huanu, which they translated as guano, back to their native lands for scientific study. They found it contained many different elements and was particularly rich in nitrogen, in the form of urea, but saw no value in it. Only later, when farmers spread the Chinchas' guano on their fields and saw crop yields increase dramatically, did they realize its potential. The guano rush was on.

In the 1850s the Chinchas became the most valuable real estate in the world. Centuries of waste from billions of seabirds had collected on the islands. The weather conditions had dried it out and concentrated the nitrogen compounds within it. These tiny islands, totalling less than two square miles, were buried ten stories deep in the richest fertilizer on Earth. Hundreds of Chinese labourers were forced to work in appalling conditions digging out the guano, wheeling it to the cliff edge and tipping it into canvas chutes that dropped straight into the holds of ships anchored below. A hundred more ships floated in the surrounding sea waiting their turn.

In 1863, Spain laid claim to the valuable islands as payment for debts Peru had incurred during the War of Independence. Peru, supported by Chile, declared war. Spain was defeated in the naval battles that followed and Latin America's true independence from Spain was sealed. But the guano boom was coming to an end and Peru faced financial ruin. By 1877 the Chinchas had been stripped down to bare rock. Eleven million tonnes of guano had been exported, farmers had become dependent on fertilizer and the race to find a new source of nitrogen began again.

Oxygen
The Maverick

Oxygen is a radical, a maverick, a rebel looking to cause havoc at any opportunity. The element's exuberance gives us life, but that comes at a price. The cost of using oxygen as a source of energy is a heavy investment in control measures and damage limitation.

The eighth element in the periodic table cannot be blamed for its fiery nature, it is in its make-up. Each oxygen atom has six electrons available to interact with electrons on other atoms. Two oxygen atoms can pair up to share their electrons and form a stable molecule, O_2. This is the most common form of the element on Earth, a colourless, tasteless gas that makes up just over 20 per cent of the air we breathe. But the way the two atoms bond to each other makes this molecule very uncommon. The twelve electrons surrounding the two oxygen atoms are paired up and distributed in such a way as to leave one lonely electron on each atom.

Molecules with unpaired electrons are called radicals, and with good reason. Though the name was originally derived from the Latin *radix*, meaning 'root', changes in the original chemical theory mean that radicals are now associated with the more modern definition of the word: instability or extremes. Electrons hate being on their own and will go

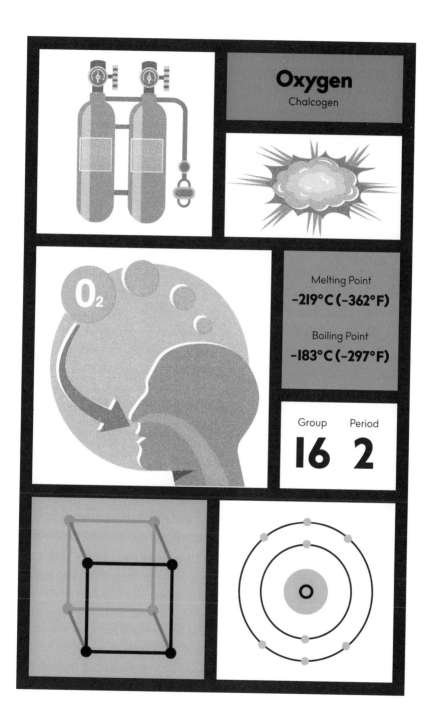

Oxygen
Chalcogen

Melting Point
-219°C (-362°F)

Boiling Point
-183°C (-297°F)

Group
16

Period
2

out of their way to find a partner to the extent that they will kidnap electrons from other molecules. Some radicals are more stable than others. Some will react with the first thing they bump into, others are choosy about the molecules they take electrons from. Molecules with two unpaired electrons are called diradicals and are usually too reactive to isolate. O_2 is a relatively stable diradical, however, meaning we can be surrounded by it in Earth's atmosphere without spontaneously combusting. But diradical oxygen is still reactive enough to combine with the food we have eaten to give us energy, with dyes and pigments to make them fade, and with metals to make them rust.

It is only through photosynthesis that such a reactive element can be so abundant in our atmosphere. Without plants, bacteria and algae replenishing stocks for us to breathe, all the oxygen in the atmosphere would have reacted with elements in the earth long ago and life on this planet would have taken a very different evolutionary path.

Animals and plants rely on the reaction between oxygen and glucose in respiration to release energy easily and abundantly, but you can have too much of a good thing. Respiration processes involve careful corralling of O_2. The gas needs to be kept out of places it should not go and emergency measures must be available to limit the damage when things go wrong. Oxygen absorbed into the body through the lungs would cause untold havoc if left to roam free. Instead, it is shepherded to specific locations, the mitochondria, where elaborate sequences of chemical reactions, controlled by a catalogue of enzymes, manage the slow step-by-step release of oxygen's chemical energy.

This careful chaperoning of O_2 on its way to the mitochondria is essential. The most dangerous thing that can happen to O_2 is that one of the unpaired electrons on the molecule gains a partner. By picking up a spare electron, oxygen becomes superoxide – literally oxygen but with extra powers – and as any comic will tell you, with great power comes great responsibility.

By reacting with water, superoxide can sire a whole series of reactive oxygen species (ROS). On the one hand ROS are vital; the body has uses for them, but in very specific roles, such as signalling between cells and keeping the body on an even keel. They can also be produced in huge quantities by the immune system to destroy invading micro-organisms. On the other hand, living things that rely on oxygen must also protect themselves from their own creation. Checks and balances are needed to prevent the immune system over-reacting. Superoxide used to destroy bacteria cannot be allowed to destroy things the body needs. Enzymes must be built to patrol the body and destroy ROS that have gone rogue, otherwise their highly reactive nature could damage fats, proteins, cell membranes or DNA. Structures must be checked, and damage repaired. Relying on such a reactive molecule to give us energy can be a double-edged sword.

9

F

Fluorine
The Great Destroyer

Fluorine is a truly terrifying element. Like a comic book supervillain, it seems hell-bent on destroying everything in its path. A villain so dangerous that, when it is finally captured, it must be contained in a specialized prison that can resist its destructive superpowers.

Like many supervillains, fluorine's violent behaviour is driven by greed. Fortunately, this element does not want world domination, the contents of Fort Knox or the destruction of the entire human race. All fluorine craves is electrons, specifically one extra electron per atom to complete its outermost shell. It might not be much to ask, but the tantrum fluorine throws to get its own way can be spectacular. Such unreasonable behaviour may seem out of all proportion for a tiny request, but most of the time it is best not to argue with this element and just hand over the goods.

All chemistry is driven by the desire of atoms to have a complete set, or at least a neatly arranged group of electrons in their outer shell. Fluorine atoms have seven electrons already, just one short of a complete set of eight. All the elements in fluorine's family, known as the halogens, face the same problem, but none of them go about acquiring their extra electron quite so aggressively and destructively as fluorine.

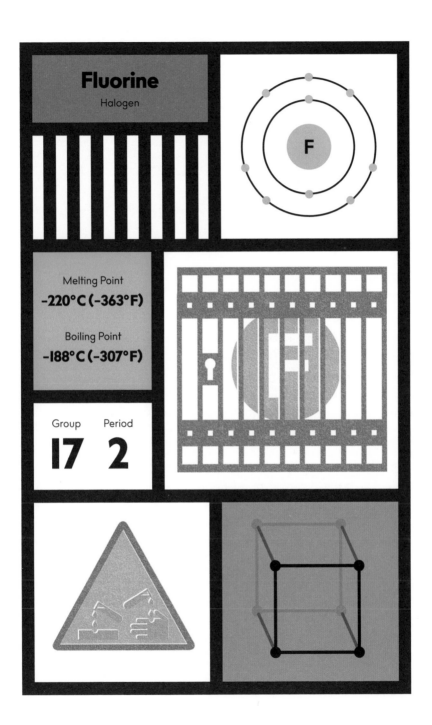

Fluorine

Halogen

Melting Point
-220°C (-363°F)

Boiling Point
-188°C (-307°F)

Group
17

Period
2

This has a lot to do with fluorine's size.

Fluorine atoms are very small, certainly compared to their fellow halogens. The outer shell of negative electrons is therefore relatively close to fluorine's positively charged central nucleus. The closeness of the positive charge helps to draw an extra negative electron into the outer shell and hold it tight once it is there.

In the absence of any freely available electrons, fluorine atoms will react with themselves as a temporary measure. Two fluorine atoms will each share one of its electrons with the other so they can kid themselves that each has a full set. Being identical twins, neither of the fluorine atoms has any extra pull, or weakness, compared to the other one, which would allow the complete transfer of the electron. But with this being the most acquisitive element in the periodic table, very few others can defend themselves from fluorine's thieving nature. If there are other elements around, a fluorine atom will quickly abandon its twin and go after the more readily available electrons elsewhere. Fluorine is consequently very, very reactive and is never found in its pure state in nature.

In the nineteenth century, scientists speculated that some minerals might contain a previously undiscovered element. There was also a convenient gap in the halogen family and fluorite became a candidate as the host for the missing element. Scientists also suspected that the missing element would be very reactive. Those who tried to hunt down pure fluorine soon learned just how reactive it is, for skin, glass and most metals are all chewed up by the element as it scavenges electrons – the walls of any prison built to contain pure fluorine must be made from elements that can resist its corrosive powers. And so, despite many valiant attempts, fluorine managed to evade capture for decades. Though nineteenth-century scientists took all the precautions they thought necessary, many were caught by surprise at fluorine's frightening behaviour. Several were left bedridden by fluorine poisoning or blinded by fluorine explosions. Some were killed.

It wasn't until 26 June 1886, seventy-four years after the first disastrous attempts, that the French chemist Henri Moissan became the first person to isolate the element. Moissan had constructed specialist equipment using platinum containers with fluorite windows, to observe pure, pale-yellow fluorine gas. Nowadays, most people avoid working with pure fluorine if they can help it and no one even tries to store it for any length of time. If a chemical reaction absolutely demands the use of fluorine gas, it is made on the spot and reacted immediately.

Thankfully, once fluorine has what it wants, it becomes much happier. Fluorine atoms that have completed their outer shell of electrons, by outright stealing or when sharing with another element, can suddenly become very cooperative. Fluoride, a fluorine atom with an extra electron all to itself, is so safe you can smear it on your teeth to strengthen them. Electrons shared between fluorine and carbon make these bonds stable enough for the compounds to be used as the non-stick coating in frying pans, or anaesthetics in the operating theatre.

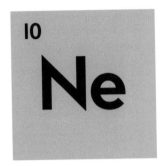

Neon
The Bright Light

Elements have been discovered in many different ways, some more spectacular than others. Previously unknown elements have revealed themselves by showing off some distinct feature – for example, a lump of metal that was unusually heavy or an unexpected reaction in a test tube. Neon, a rare, unreactive gas that had gone completely unnoticed until 1898, had a particularly grand unveiling.

While some elements were discovered by accident, many were hunted out deliberately. In the late nineteenth century, scientists made use of the unique pattern of light emitted by each element when heated to identify the element and seek out new ones. A sample, a spectroscope and a library of spectra for comparison were all that were needed. A tiny band of colour that did not appear in the spectra of any other known element was enough to get an element hunter's pulse racing.

In 1895 Sir William Ramsay and Morris W Travers were carefully separating out the components of ordinary air. Each component, be it element or compound, has a unique melting and boiling point. So, after condensing air into liquids and solids, the temperature could slowly be raised to boil off each component separately. Every time they thought they had isolated a unique gas, the scientists passed a current

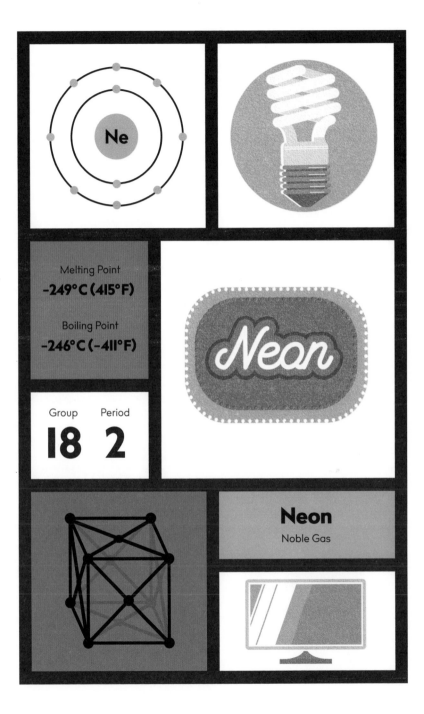

Ne

Melting Point
-249°C (415°F)

Boiling Point
-246°C (-411°F)

Group 18 Period 2

Neon

Neon
Noble Gas

through it. They then peered at the light it emitted through their spectroscope in search of new bands of colour.

Their painstaking work paid off. Ramsey and Travers isolated not just one element but discovered a whole family – the noble gases. First came argon (meaning 'lazy'), then krypton (meaning 'hidden'). After spending hours and hours refining huge volumes of gas, over and over again, Ramsey and Travers collected ever smaller fractions of ordinary air. With scarcely a thimbleful of gas from their latest round of refining, they passed a current through it expectantly. This time there was no need to reach for their spectroscope, the tube glowed bright red. Ramsey and Travers were temporarily struck dumb by the bright light. This gas glowed like no other. It was decided to call the element neon, after the Greek *neos* for new, a name it more than lived up to.

Neon, like its fellow noble gases, has a perfect arrangement of electrons around its nucleus meaning it is not interested in sharing or exchanging them with other atoms to form compounds. Like its sister element helium, neon may never have formed a chemical bond in its entire existence. The effort of removing electrons from a neon atom requires more energy than a chemical bond can provide. But there is more than one way to skin an atom of neon.

By reducing the pressure, and applying a few thousand volts of electricity, neon can be persuaded to part with some of its electrons. When the negatively charged electrons are wrenched away, they leave behind positively charged neon atoms. The negative electrons are attracted to the positive neons and they recombine releasing the energy that was needed to pull them apart. As long as the high voltage is applied, electrons will be pulled away and recombined with the parent atoms over and over again. The result is the bright red light Ramsey and Travers first saw in their laboratory.

The substance created by positively charged neons floating in a sea of negative electrons did not conform to any of the traditional states of matter: solid, liquid or gas. It reminded the scientist Irving Langmuir of how red and white blood cells flowed around the body in a special fluid, so, in 1928, he named the new state of matter 'plasma'. This new state of matter led to several technological advances, from fluorescent lamps to flat-screen TVs.

For a large part of the twentieth century, neon's red glow symbolized modernity. The glass tubes that contained the gas could be bent and twisted into extravagant shapes. Neon's light was vivid in colour and bright enough to be seen in dazzling Californian sunshine or through thick London smog. And all available at the flick of a switch. It was perfect for advertising.

The first neon advertising sign shone out on the Champs-Élysées in 1913. A decade later they started appearing across the United States. Soon they were everywhere, epitomising the brash decadence and cheerful optimism of the era. Then, just as neon signs were becoming passé, neon found a new application. In 1960 the first visible laser beam shone in a powerful straight line that could reach the moon. The laser's red glow was from neon.

Sodium
The Silent Partner

Sodium is the solid, reliable workhorse of the periodic table. It is seemingly everywhere in our lives, from nerves to nuclear reactors. Its ordinariness may appear mundane, but it should be celebrated for fitting into so many different roles, performing them well and with so little fuss. Sodium's special talent is to be almost omnipresent while remaining unobtrusive. It facilitates a lot of important processes but rarely takes centre stage – like the straight man in a comedy double act, absolutely necessary for the joke to work, but rarely getting to deliver the punchline.

The enthusiasm for working in partnerships with other elements is sodium's strength. Take cake-baking as an example. It is the baking soda, or sodium bicarbonate, that adds the sponge to Victoria. Tiny bubbles of gas are released from the carbonate bit of the molecule during the baking process. The sodium adds nothing other than a convenient way to introduce the bubble-making component into the mixture. It is all down to sodium making salts, and not just the kind you sprinkle on your chips.

Each atom of sodium is desperate to lose the single electron in its outer shell because it leaves behind a neat and complete shell of

Sodium

Alkali Metal

Melting Point
98°C (208°F)

Boiling Point
883°C (1621°F)

Group   Period
1   **3**

Na

electrons underneath. With one less negative electron, sodium becomes positively charged Na^+. The electron it has shed is not left roaming around unwanted; there are always plenty of other elements that are willing to take up any spare electrons to complete gaps in their own outer shells. One such element is chlorine, and it uses sodium's unwanted electron to form chloride, Cl^-. The attraction of positive and negative between Na^+ and Cl^- holds them together in crystals, like alternating bricks in a vast, regular three-dimensional block. This is common, table, rock or sea salt.

Confusingly, the word salt also refers to any molecule with a distinct negative and positive component. Salts, especially sodium salts, have qualities that make them extremely useful in our daily lives. They are solids, making them convenient to store, but they dissolve in water so the positive and negative parts can be distributed. After we eat food seasoned with table salt, the Na^+ and Cl^- are separated in our watery bodies and each can go off and do its own thing. For the sodium half, it goes on to form yet another partnership, this time with potassium, a double act so important that it keeps us alive.

Sodium and potassium are from the same periodic family: the alkali metals. They are similar to each other, but it is the differences *between* them that count. Their coordinated dance in and out of nerve cells generates the electrical pulse that sends signals around the body to keep us on an even keel.

The membrane that surrounds all nerve cells contains channels that allow specific elements and molecules in and out; there are potassium channels for potassium, and sodium channels for sodium. The slight differences between the two stops the wrong element sneaking through the wrong channel. Potassium drifts into the cell and sodium is pushed out before the gates are shut behind it. This chemical segregation makes the inside of the cell slightly negative and the outside slightly positive, like the plus and minus of a battery.

When the gates open, sodium rushes into the cell and potassium escapes out of it, creating the electrical pulse that runs along the length of the nerve cell, much like a Mexican wave. Not enough sodium in our diet and things can go very wrong very quickly. Our brains trick us into savouring well-seasoned food to ensure we get enough.

Though sodium salts are a familiar part of our everyday lives, sodium as a pure element is not nearly as well known. This is because, until 1807, it had never been seen on its own. Always desperate to get back into a partnership, sodium's solo career never really took off and it's most often a support act to other, showier elements. For example, inside nuclear reactors, radioactive elements are split apart to release huge amounts of heat. Pure sodium acts as the intermediary between the reactor core and the water that boils to drive the generators that give us electricity. Its lightness, low melting point and ability to conduct heat make it well suited to the task but, as usual, it is in the background doing important but unremarkable things. Sodium shows it is okay to be ordinary. Ordinary is what makes the world go round.

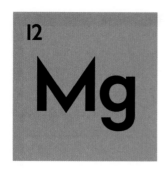

Magnesium
Our Green-Fingered Friend

In the beginning, the Earth was a barren and hostile place. The transformation from rocky wasteland to the lush green home we enjoy today is down to chlorophyll. This magnesium-centred molecule has facilitated the metamorphosis of our planet. Magnesium has been the midwife to the diverse range of life we see all around us.

All living organisms require energy, and in a form that can be used to carry out the necessary chemical reactions that make life possible. A single-celled organism can get by with pretty meagre rations. A whole host of exotic compounds and elements can be manipulated to release scraps of energy. But, if you want something more exciting than bacteria or algae for company you need oxygen, and a constant supply of it.

Oxygen reacts with other elements easily and releases a lot of energy in the process. But once reacted, oxygen is a spent force. It must be pulled away from its chemical partners, so it can react again. When life first appeared on this planet all the oxygen had already been locked up in compounds. It had done what oxygen does best and reacted with the metals, carbon and rocks that formed the Earth itself. Plenty of oxygen was sloshing about in water but prising it free was beyond the resources of these early organisms.

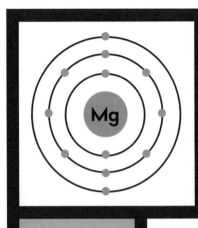

Group
2

Period
3

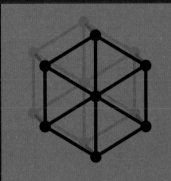

Magnesium
Alkaline Earth Metal

Then, at some point in our evolutionary past, a single-celled organism produced a bright-green compound. It changed life on Earth. Chlorophyll allowed plants, algae and some bacteria to harness energy from the sun into chemical reactions. It was not the first compound to do this, but it was the most efficient. Most importantly, unlike other light-harvesting processes, it could split water to release oxygen.

Chlorophyll is an elaborate arrangement of carbon, hydrogen, nitrogen and a few oxygen atoms. But it would be a relatively dull molecule without a single atom of magnesium at its heart. Evolution has tried other metals, but a magnesium-based chlorophyll system turns out to be the most efficient.

The chlorophyll molecule works like a solar panel, collecting light energy and funnelling it into the chemical reactions of photosynthesis. An intensely coloured molecule is needed to absorb visible light. The green colour produced when magnesium slots into place inside a chlorophyll molecule is ideal. It absorbs red and blue light very strongly; a compromise between harnessing the most energy and getting the most reliable supply. Absorbing more green light would gain more energy, but it would be more disrupted by shadows passing overhead.

Once chlorophyll had been made, life on Earth never looked back. Organisms that could tolerate the highly reactive oxygen that suddenly surrounded them, thrived. Chlorophyll cleared the path from single-celled organisms to everything else; from trees to *Triceratops* to us.

It took millions of years and billions of tiny evolutionary steps to get to where we are today, and the journey has not always been easy. The oxygen levels we enjoy now were not built up in a day. There have been peaks and troughs. Life is a constant cycle of oxygen being locked up and released again by photosynthesis.

Evolution has only tightened the ties that bind us to chlorophyll. Humans are utterly reliant on this molecule and in more ways than one. We need the oxygen it provides, but also the magnesium it contains. Inside our bodies, magnesium maintains our bone structure; it helps build proteins and replicate our DNA, among many other fundamental processes that keep us alive and healthy. All our magnesium comes from the chlorophyll in the plants we eat, and the plants eaten by the animals we consume.

Though we may not be conscious of it, we have evolved to appreciate chlorophyll. The world appears to us much greener than it really is. Our eyes are particularly well tuned to enhance the green appearance of our environment. We also notice when it is gone. At the end of summer, when the days become shorter, there is less light available for photosynthesis. Many species of tree dismantle their chlorophyll and pack away the pieces for the winter. Magnesium is removed from its home at the centre of the molecule and the green colour disappears. People will travel from far and wide to marvel at what is left behind. The de-greening of the foliage reveals other pigments that have been helping chlorophyll absorb the best light all summer. Take away magnesium and you are left with the yellows, reds and oranges of autumn.

13
Al

Aluminium
The Lightweight

The guests who sat down at the emperor's banquet must have been slightly perturbed. Gone were the gold knives and forks, the usual signs of wealth and power. In their place were what probably looked like badly polished silver cutlery. This tableware had none of the reassuring weight and solidity of gold, or even silver and, still worse, it made the food taste odd. Was this some kind of elaborate joke? The emperor told his guests they were honoured to be using such strange dining implements because they were made from a metal so rare it was more precious than gold. And, because the emperor had said so, so it was.

Emperor Napoleon III of France really did have his table laid with aluminium cutlery to impress his most esteemed visitors. This was not a ruse to trick the emperor or his guests. In the mid-nineteenth century, aluminium was the latest thing and really was more valuable than gold. Using it for forks and spoons was incredibly ostentatious, even though it made for a slightly disappointing dining experience.

Alum, from the Latin *alumen* meaning 'bitter', has been known about for millennia and used to fix dyes to fabric. Suspicions that alum was a salt of some kind, a compound of a metal and non-metal, began in the sixteenth century. By the mid-eighteenth century, concerted

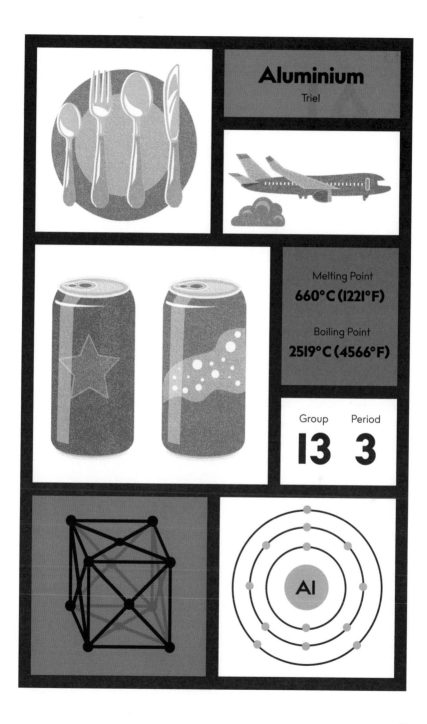

Aluminium

Triel

Melting Point
660°C (1221°F)

Boiling Point
2519°C (4566°F)

Group Period
13 **3**

Al

efforts were being made to isolate aluminium from alum and similar salts. The problem was aluminium's high reactivity. It was extremely reluctant to detach itself from its non-metal partners.

Eighty years later, Hans Christian Ørsted and Friedrich Wöhler, each working independently, succeeded. Their chemical methods of separation relied on offering a more reactive metal to lure away the non-metal components of the salts and leave aluminium behind. These reactions were unreliable and often left impurities in the aluminium. Obtaining more reactive metals was also a challenge. Potassium and sodium were reactive enough, but to get pure potassium or sodium required electrolysis, a costly and inefficient process at the time. The resulting scarcity of aluminium made it prohibitively expensive.

Aluminium's reactivity has advantages as well as disadvantages. Its particular affinity to oxygen means the surface of aluminium oxidizes quickly, but unlike other metals, where reactions with oxygen cause corrosion, aluminium oxide forms a protective layer that maintains the lustre of the underlying metal. But this protective coat is not perfect. The acids found in many fruits and vegetables can corrode the oxide layer to attack the underlying metal and taint the flavour of the food. Napoleon's gold cutlery would have delivered food to his guest's mouth unreacted and with optimum flavour, but it just was not as impressive.

The French emperor was not the only one to value aluminium so highly. In the United States, when designing the monument to commemorate the country's first president, it was decided it should be topped off with something that demonstrated the country's wealth and importance. The Washington Monument was duly capped with a 2.7-kg (6-lb) pyramid of aluminium. The 22-cm (9-in) -high pinnacle

was presented to the general public in a special exhibition held at Tiffany's. The monument's cornerstone was laid in 1848 when aluminium was exceptionally difficult to get hold of and consequently at its most expensive. By the time the monument was opened to the public in 1888, however, aluminium prices had plummeted.

The reason for aluminium's fall was industrialization. Processes for refining aluminium ore and extracting the metal it contained using electricity were developed in the 1880s. Pure aluminium could now be mass produced. This lightweight but strong metal was ideal for planes, trains and automobiles. The two world wars caused demand for aluminium to skyrocket. By 1945, infrastructure was in place to produce vast quantities of the metal but, post-war, it needed more commercial applications. Soon it was used in everything from toys to saucepans.

In less than a hundred years aluminium had gone from regal to regular. Now everyone could afford to set their table fit for an emperor, but no one did. Eating food with knives that could not keep a sharp edge and spoons that spoiled the flavour could only be some kind of joke.

14

Si

Silicon
The Hybrid

If there is one element suited to science fiction, it is silicon. This weird hybrid, between organic and artificial, biological and electronic, has all the properties that lend themselves to fantastic structures and strange behaviour. It is understandable that science-fiction writers speculating about unusual life forms often mention it.

Silicon's position in the periodic table tells you much about its contrasting nature. It sits at the junction between two very different classes of elements. To one side are the elements we associate with the abundance of organic life all around us. And, to the other, are some of those elements humans have manipulated and incorporated into their own synthetic world of technology. Silicon has a foot in both camps. It has biological roles and also enables the digital world we have built.

It is silicon's similarity to its chemical sibling carbon that suggests it would be an ideal substitute in the staggering number of biological functions usually carried out by carbon. It is present inside our bodies, particularly in hair, bone and collagen, but whether it is essential and what the optimum daily intake might be is hard to establish. Silicon is so abundant in our world – far more so than carbon – that studying silicon deficiencies in our diet is impossible. It also raises the question

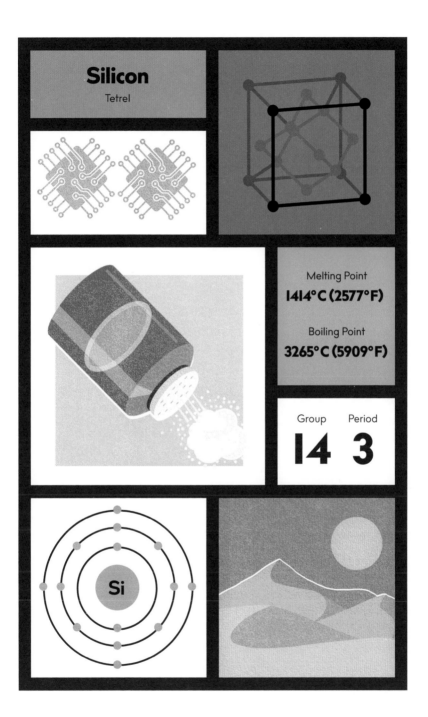

Silicon

Tetrel

Melting Point
1414°C (2577°F)

Boiling Point
3265°C (5909°F)

Group
14

Period
3

Si

that, if silicon is so abundant and chemically similar to carbon, why is it not a bigger part of our biological lives?

Carbon has certain advantages over silicon. It forms a mind-bogglingly huge range of molecules containing anything from two atoms to thousands in seemingly endless combinations and arrangements. Silicon can also form small and very large molecules, certainly on a biologically useful scale, but with considerably less variety. If carbon is the basis of complex symphonies of molecules, silicon has a repertoire of simple tunes picked out on a piano.

The other important factor for life is how easily chemical bonds can be broken and remade for life to grow and develop. For carbon, the strength of its chemical bonds is within a goldilocks range that allows easy rearrangement and redistribution of atoms among its many different molecules. This flexibility enables biological processes to be carried out in the moderate conditions found on this planet. Silicon cannot quite match carbon in terms of the amounts of energy stored and released by the breaking and making of its chemical bonds. In particular, silicon forms weak bonds with itself and strong bonds with oxygen. The result is that most silicon on Earth is bound up as silicates, or sand.

You might think that silicon-based biological systems would come to a juddering halt when the silicon is locked away in chemically inert sand, but instead it presents a fantastic opportunity. The compounds silicon forms with oxygen represent one of the few areas in which it surpasses carbon's precocious chemical abilities. The beauty and diversity of silicate structures range from ordinary glass to garnets and from talcum powder to topaz. Silicates are abundant, strong and versatile and they have been used by some of the most exotic looking creatures on Earth. Who needs science fiction when science fact conjures underwater opals and glass sponges?

Diatoms, a kind of marine algae, absorb the tiny amounts of silicate dissolved in seawater and use it to encase themselves in protective opal coats. The complex crystal arrangement of silica and water in these microscopic shells is the same as is found in the opals valued as jewels. It is why both reflect light in an iridescent spectrum of colours.

Then there are the glass sponges growing on reefs hundreds of metres beneath the waves. These strange creatures do not have cells like us and most other life forms on Earth. Their internal network of channels and pores are formed from silicates spun into a fine framework. These rigid underwater glass clouds cannot move. They let life drift past and through the tiny gaps in their delicate structure to filter out the micro-organisms on which they feed. A life based around silicates may be simpler and slower, but it is no less spectacular than any other, and it is not the only possibility.

Freed from oxygen, silicon can tap into the other side of its hybrid nature. Pure silicon has some similarities with its neighbouring metals, in that it can conduct electricity, but only in certain circumstances. This means the flow of electricity can be controlled; decisions can be made, and information transmitted. It might be yet another form of life, just not as we know it.

15

P

Phosphorus
The Gothic One

Plenty of elements have their dark side. The list of those that have
been misused for murder or warfare is long, and phosphorus is no
exception. But its association with stories of alchemists and ghosts
sets this element apart from others. Phosphorus is not just dark;
it is positively Gothic.

It all started with Hennig Brand, a seventeenth-century alchemist
who searched for gold in his urine. After weeks of stockpiling, boiling and refining his personal commodity he peered expectantly into a
chemical flask in the hope of spotting a faint glimmer of gold at the
bottom. Instead, he saw a white solid that glowed with a green light.
This new substance was named phosphorus meaning 'light bearer'.

Brand went back to his gold studies, but other scientists became
obsessed with the mysterious glowing phosphorus. It was smeared
on hands and faces to create spooky effects in scientific salons.
Words were written on pieces of paper using sticks of phosphorus
and when the corner of the word was scratched it would inexplicably
burst into flames. This fantastic demonstration drew the crowds to
the increasingly popular public science lectures that were appearing
all over Enlightenment Europe.

Phosphorus

Pnictogen

P

Melting Point
44°C (111°F)

Boiling Point
281°C (534°F)

Group **15** Period **3**

People marvelled at the strange properties of phosphorus, but the causes of the ghostly glow and frightening flammability were a puzzle. The truth was not discovered until centuries later, and when it was found it solved a few other mysteries as well.

Like many Gothic villains, phosphorus has a defining obsession with one thing above all others, and in phosphorus's case it is oxygen. Given the slightest chance, phosphorus will form very strong bonds with oxygen atoms, releasing a lot of energy as a result. If the oxygen is introduced in a calm, measured way, the energy is given off as a greenish light, with no heat, in a process called chemiluminescence. But, with very little provocation, such as the heat of friction from a scratch, phosphorus will grab all the oxygen it can to produce a lot of heat, bright flames and choking white smoke.

From the moment of its discovery by Brand, people knew that phosphorus was within our bodies, but how could we survive having such a temperamental element within us? Especially given the amount of oxygen we draw into our bodies with every breath. The answer is that once phosphorus has all the oxygen it wants, it becomes stable, content phosphate. An invaluable structural unit within all living things, phosphate forms the backbone of DNA and is involved in fundamental processes within almost every cell in every living thing.

Anything an animal eats that has ever lived, be it animal or vegetable, will introduce more phosphate into its system. There can be so much that the excess has to be expelled. Rather than being wasted, this is an opportunity for micro-organisms to get their share of phosphate. These bacteria and other microbes get their food and energy from life's leftovers. And, as any good Gothic story will tell you, where there is life there is death.

In forests, marshes and graveyards, mounds of dead material build up. In true Gothic style, such environments produce tales of will-o'-the-wisps, jack-o'-lanterns or similar spooky lights flitting through the trees or hovering over marshes like phantoms. Stories of these phenomena have been around for a long time but are usually dismissed as supernatural nonsense. Science, however, may have an answer, or at least a plausible theory to explain the phenomenon.

Some forms of life owe their existence to decay. They are essential for recycling phosphorus and other elements back into the earth to support new life and growth. Some of these decay-loving life forms have the power to prise phosphorus away from its beloved oxygen to make diphosphane, a gas that spontaneously catches fire when it comes into contact with the air. Where there is a lot of dead material these micro-organisms of decomposition are treated to a banquet. The by-products of this bacterial feasting accumulate. Occasionally these by-products are belched out as clouds of methane, or marsh gas, and, for the most part, they float away unobserved. But, perhaps, every now and then, diphosphane is also released. Suddenly reintroduced to the oxygen they have been deprived of by the bacteria under the soil, these molecules might glow, or even ignite the methane around them.

So, if you go down to the woods tonight, you are in for a big surprise. If you go down to the woods tonight you might see phosphorus in disguise.

Sulphur
The Acid Rain Maker

Many elements hide the darker aspects of their nature. Elements that cause harm usually do so using subtle and insidious methods. By contrast, sulphur has never tried to disguise its awfulness. It almost brags of its evil associations and delights in its connections with the devil. Sulphur's embarrassing secret is that it also does a lot of good.

Sulphur is one of the few elements that have been known since antiquity. It probably was not discovered; more likely a point was reached where it could no longer be ignored. Pure sulphur emerges from hot springs and volcanoes, accompanied by the stench of rot and decay.

Sulphur is at the heart of many malodorous molecules. It is the key to garlic's pungency and the powerful stink of methanethiol, which alerts us to gas leaks. Compounds such as hydrogen sulphide and dimethyldisulphide are part of the stench of death and decay. Such olfactory warnings guide some animals like us away from food that may be past its best.

Back at the volcano, the brave few who were not put off by the sulphurous smell and burning temperatures would have been rewarded with quite a spectacle. The first people to clap eyes on pure sulphur must have been surprised to see bright-yellow rocks that melted into a

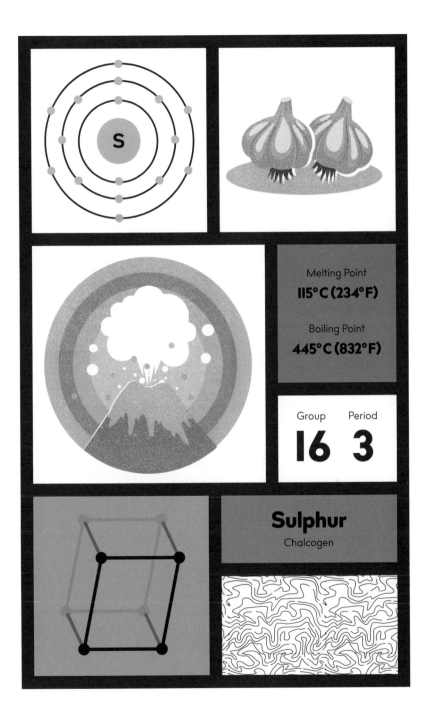

Melting Point
115°C (234°F)

Boiling Point
445°C (832°F)

Group
16

Period
3

Sulphur
Chalcogen

blood-red liquid and burned with eerie blue flames. Given the setting and appearance of this element, it cannot have been much of a stretch to believe it had been belched out of hell itself.

This fearful element became the brimstone, literally 'burning stone', of the Old Testament. The threat of burning sulphur has been used to intimidate people for thousands of years. Sinners were scared into mending their ways with stories of souls condemned to eternity in the fire and brimstone of the next world. Sulphur also threatened physical bodies in this world. Combined with carbon and saltpetre it makes gunpowder.

The environment was not left unscathed by sulphur's malign presence either. Humans have made something of a Faustian pact with fire and sulphur. Many of our technological advances have been fuelled by the human ability to control fire. As our population and technical ability have grown, so has our demand for fuel to power those advances. Fossil fuels, rich in sulphur compounds, have been exploited to meet that demand. The dark satanic mills of the Industrial Revolution were powered by coal that billowed clouds of sulphur dioxide into the atmosphere. Cities choked on the smoke from open fires and factory chimneys. The oil products burned in engines that transported us around for much of the twentieth century released yet more sulphur dioxide into the air. There, it dissolved in the water droplets in clouds and returned to Earth as acid rain.

In keeping with its devilish reputation, sulphur does not just damage and destroy, it also corrupts. After spending centuries in the company of sulphur, even (stone) angels can acquire dirty faces. Sulphur has a strong attraction for certain metals. Old bibles, sumptuously illustrated with holy images, have been slowly tarnished by sulphur released from fires bonding to the lead in the white lead paint. Over time black-brown lead sulphide smudges have appeared on the cheeks of even the saintliest figures. Sulphur is now stripped from factory chimneys and refined out of crude oil. Not only is the atmosphere better for it, but catalysts, made from heavy metals and used in industry as well as car exhausts, are no longer coated and poisoned by sulphur's presence.

It is easy to demonize an element responsible for rotten smells, gunpowder and acid rain. But alongside the bad and the ugly, there is also the good. The sulphur in fossil fuels is no geological accident, it is there because these fuels are made from once-living things. Sulphur is as much a part of life as death. Atoms of sulphur are sprinkled along the strings of amino acids that form proteins and enzymes. They pair up like molecular Velcro and are strong enough to hold the amino-acid thread in carefully wound, three-dimensional shapes that allow the protein or enzyme to function correctly. But they are also weak enough to be pulled apart when the enzyme needs to be switched off. Keratin, the structural component of hair and nails, contains a lot of sulphur. Whether through the stories of hellfire or the science of sulphur bonds, sulphur can literally make your hair curl.

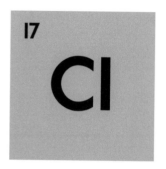

17

Cl

Chlorine
A Split Personality

Chlorine is the Jekyll and Hyde of the periodic table. Many people know its malevolent side from reports of its destructive effect on ozone and accounts of its use during the Great War. It is easy to forget that chlorine, as its alter-ego chloride, is essential in order for life to thrive.

There is no doubt that chlorine has its vicious side. Like its sibling, fluorine, it desperately wants an extra electron to complete its outermost shell and will try to grab one from anything it can. Few elements can hold out against fluorine's violent attacks, and chlorine is only slightly less intimidating. But grabbing that extra electron is as transformative as Dr Jekyll's serum. The violent, amoral chlorine/Hyde character is completely repressed, and the mild-mannered, popular chloride/Dr Jekyll appears.

Compounds of chlorine, such as sodium chloride (or salt) and hydrochloric acid, have been known about for a long time. However, it was Swedish chemist Carl Wilhelm Scheele who first isolated the element and systematically studied it in 1774. His work revealed important characteristics of the sickly, pale, green-yellow gas. It reacted with almost everything it came into contact with, had a choking smell, bleached paper, plants and cloth, and killed insects.

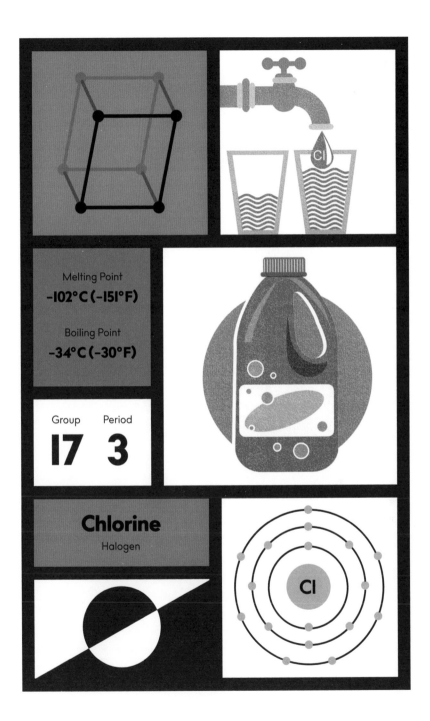

Melting Point
-102°C (-151°F)

Boiling Point
-34°C (-30°F)

Group
17

Period
3

Chlorine
Halogen

Cl

It took a few years for people to catch on to the bleaching potential of chlorine, but when they did, it transformed the paper and cloth industries. Further improvements were made by bubbling chlorine gas through a solution of sodium hydroxide. It made sodium hypochlorite – what most people call bleach. Liquid bleach was easier to work with than choking chlorine gas. And the bright white cloth it produced did not just look clean, it was clean.

Bacteria and viruses are particularly sensitive to hypochlorite, even in very dilute solutions. Bleach is not a medicine, but hospitals can be scrubbed down with it, slashing mortality rates. In 1897, typhoid broke out in the English town of Maidstone. It was decided to try adding chlorine to the drinking water and an epidemic was averted. Most of the developed world now uses chlorination to treat its drinking water. The quantities are far too small to affect the health of the people drinking the water, but it destroys all manner of water-borne pathogens that have claimed the lives of millions of people in previous centuries.

But the darker side of chlorine's nature has never been far away. On 22 April 1915 clouds of chlorine gas were released from cylinders along the Western Front at Ypres. A gentle breeze pushed the clouds towards the opposing trenches. The gas, heavier than air, seeped across the mud of no-man's land, settled into hollows and poured into dugouts.

When the chlorine reached the moisture in the soldiers' eyes, mouths and lungs it was converted into hydrochloric acid. It was the first time chemicals had been chosen, manufactured and deliberately released in a war setting with the full knowledge of the likely consequences. Thousands died and thousands more were left blinded, or with wrecked lungs. This horrific development shocked the world and at the end of the war new international resolutions were drafted to ensure that nothing like that could ever happen again. They were not entirely successful.

Dr Jekyll found that once Mr Hyde was released, he was increasingly difficult to control. As the twentieth century progressed chlorine came to be seen as the most polluting element of all. The chlorine-based chemicals increasingly used in plastics, pesticides, refrigerants and aerosol propellants were finding their way into the Earth's upper atmosphere. There, ultraviolet light breaks these molecules apart to release chlorine radicals, highly reactive atoms that rip through ozone.

As evidence of the dangers posed by a hole in the ozone layer became apparent, many synthetic chlorine compounds have been either banned or heavily regulated to curb the damage. In 1991, out of environmental concerns, Peru decided to discontinue the chlorination of its water supply in many of its regions. The result was a massive cholera outbreak that claimed the lives of 10,000 citizens. Chlorination was reinstated.

There are at least two sides to every story. During the Great War, chlorinated drinking water and hypochlorite solutions, for treating open wounds, saved many lives. Chlorine-based chemicals are polluting, but they are also essential for life; from the acid in our stomach to the chloride regulating our cells. As for Dr Jekyll, the good and the bad are not always so easy to separate.

19

K

Potassium
The Showman

Potassium is extremely stubborn and likes a quiet life – prizing it out of obscurity requires much persuasion. And don't expect this element to be grateful to you for dragging it out into the light. Those who dare to isolate potassium and fail to treat it with the respect it deserves, soon learn from their mistakes.

Like all the other elements in the group 1 family, potassium has a single electron in its outer shell. All of its other shells of electrons are neatly filled. Moving from the head of the family and down through the group, this extra, odd, untidy layer becomes an increasingly unwanted appendage. Potassium, much more than lithium or sodium, enthusiastically donates this remainder electron to anything that will take it. Fortunately, there are plenty of elements missing an electron or two to take any that are going spare. The result is a partnership, or salt, formed between potassium and the atom that has taken on its extra electron.

This happy cohabitation is the normal and preferred way of life for potassium. The conditions on planet Earth mean that this element is never found naturally in its pure state. Scientists in the early nineteenth century were familiar with potassium salts such as potash, and suspected that it was not a pure element. However, no one had succeeded in

Potassium
Alkali Metal

Melting Point
64°C (146°F)

Boiling Point
759°C (1398°F)

Group   Period
 1   4

K

separating it from the other elements potash contained. Then, a scientist desperate to have something to show off, dragged potassium from its comfortable existence and pushed it in front of an audience without any of its fellow elements there to support it. Potassium, as you might expect, reacted badly.

Humphry Davy started life in very humble surroundings but displayed an uncommon aptitude for science. Through a combination of ability, hard work and perseverance, he scaled his way up the ladder of the scientific establishment. His flair for public speaking and ingenious demonstrations endeared him and his work to the general public, even if the academic elite did not always share the same opinion. Nevertheless, his achievements in chemistry could not be ignored.

In 1807 Davy was invited to deliver the prestigious Bakerian Lecture at the Royal Society, home to the great and the good of the scientific world. He needed something special, something to impress even the most conservative minds attending. He decided to use the lecture to announce a great development, something that others had tried but failed to achieve. His 1807 lecture would announce the discovery of a new element. The only problem was he had just a few weeks before the lecture to discover it.

At the time, Davy worked at the Royal Institution, an establishment that had the benefits of good funding, excellent laboratory facilities and a programme of public lectures so popular they caused traffic jams in the street outside. Its facilities and financial support had allowed him to build the most powerful battery in the world. He needed it to pull

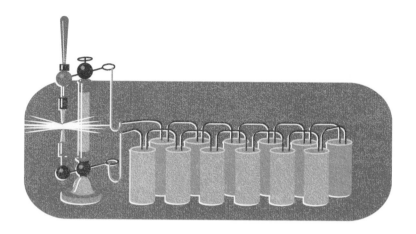

potash apart. The electrical strength of the battery forced the potassium atoms within potash to take back the electron they were normally so desperate to get rid of. Davy was the first person to lay eyes on pure potassium metal that accumulated at one of the battery's electrodes.

The heat generated by the experiment melted the potassium into globules that looked like shimmering droplets of mercury, but these droplets had a trick up their sleeves. Each potassium atom, dragged out into the open, was now encumbered with an unwanted lone electron in its outer shell. It took the first opportunity to offload the electron onto the oxygen and water in the air. Like a magician staging a dramatic disappearance, there was a burst of bright lilac flames and potassium disappeared back into potash.

When it came to the Bakerian Lecture, Davy's announcement of the discovery of a new element sparked metaphorical and actual fireworks. It was not something to which the staid academic environment of the Royal Society was accustomed. Davy's flamboyant style was not considered seemly for such serious science. Others doubted the veracity of his claims, as no one outside of the Royal Institution had the resources to replicate his experiments and confirm his findings. Fortunately, such a dramatic announcement was no flash in the pan. Today, pure potassium can be produced easily, though scientists are much more wary of its reactivity and treat it with considerable care.

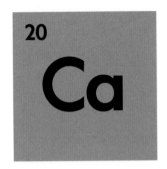

20

Ca

Calcium
The Character Actor

Many elements have distinctive attributes or standout features that equip them for a particular niche. Calcium, on the other hand, excels in many fields. It has found its way into our lives in roles as diverse as chalk and cheese. It is with us from the moment we wake up and keeps us sheltered while we sleep. Calcium is our constant companion throughout the day, holding things together so life can carry on. This solid and reliable element carries out a thousand thankless tasks, many of which go unnoticed and unappreciated.

Like a versatile character actor using make-up, costume and mannerisms to inhabit diverse roles, calcium uses chemical add-ons to transform itself into anything from a master builder, to family physician or Renaissance artist. A consummate actor can metamorphose into a wholly different personality, to the extent we scarcely recognize the person beneath the disguise. People often seem surprised that calcium, like an actor, is very different when off-screen.

It was not until the start of the nineteenth century that anyone realized calcium *was* an element, so deeply was it undercover in the guise of various compounds. An element's personality or character is defined by how its electrons are arranged around its nucleus. When elements

Calcium

Alkaline Earth Metal

Melting Point
842°C (1548°F)

Boiling Point
1484°C (2703°F)

Group
2

Period
4

react to form compounds, electrons are shared, swapped or donated. The result is much more than a blending of two sets of characteristics, it is a completely new configuration. Calcium reacting with other elements goes further than donning the most brilliant disguise, it transforms into something completely different.

Calcium is most familiar to us in the guise of rocks and bones. We often think about the element in terms of structure. It has shaped the landscape around us in chalk cliffs and limestone quarries; it is the basis of building materials that cement roads, plaster houses and prop up porticoes; and it is integral to the skeletal frame that keeps us from collapsing into a blubbery heap. But calcium is more than a passive support for our world.

While engineers and architects appreciate calcium carbonate for its solidity and permanence, they also prize it for its appearance. Designers have decorated interiors and exteriors with marble. Sculptors have rendered exquisite statues from alabaster and craftspeople have adorned objects with pearls. Such elaborations have often signalled wealth. In a grand show of decadence, Cleopatra is said to have dissolved a pearl in a cup of vinegar and drunk an impressively expensive, though revolting, calcium cocktail. Many people take calcium supplements to ensure strong teeth and bones, but the queen of Egypt perhaps took things a little far.

While bone strength is important, the skeleton is not just a rigid framework on which to hang our flesh. It is a living resource within our body, a repository of calcium that builds up and breaks down as supplies from our diet wax and wane. Ninety-nine per cent of the calcium within us is stockpiled in our bones as calcium phosphate. The remaining one per cent regulates our day-to-day lives. Withdrawals and deposits are constantly made from this bone bank to finely balance the tiny fraction of calcium that moves through our softer parts. This free-moving form of calcium causes the contraction of muscles, the release of hormones and the clotting of blood, among other vital processes.

Calcium's biological talents are carefully tailored to the situation. Strong, solid calcium phosphate is ideal as structural support in bone, but disastrous in the hectic environment found inside cells. Cell interiors abound with adenosine triphosphate. The phosphate units of this molecule are sequentially broken off to supply the energy the cells need to function. Calcium has to be rigorously excluded from such places in order to prevent the cell's frenetic way of life from grinding to a halt. Such strict policing can have unexpected benefits. The continuous pushing of calcium to the outside of cells may have resulted in the formation of hard external crusts of calcium phosphate and calcium carbonate. It may have been one of the early evolutionary steps towards shells and exoskeletons that protect many forms of marine life.

These insoluble compounds of calcium often hog the limelight, which incidentally is a brilliant white light created by a flame directed at a block of lime (calcium oxide). So, what of calcium once shorn of all accoutrements and additions? What surprises many people is that pure, undisguised calcium is a rather unremarkable metal.

22

Ti

Titanium
The Strongman

Titanium is the strong, silent type. It presents a tough exterior to the world – an armoured shell that prevents it from engaging with anything around it. But that armour is no bulky, clanking suit of dull metal. Like the suit worn by a superhero, it is close fitting, adaptable and can even be colourful. And, just like a superhero, titanium protects us humans from harm.

The element was discovered in the 1790s, found bound up in rocks with iron and oxygen. The iron was easy enough to remove but left behind an oxide that everyone was absolutely certain contained a previously unknown element. They named it titanium after the Titans, the first generation of Greek gods born of Uranus (the heavens) and Gaia (the Earth). The Titans were overthrown by the Olympians in a war lasting ten years. The chemical battle to separate titanium from titanium oxide was less brutal, but took a lot longer.

Many tried to persuade the metal to let go of the oxygen to which it was bound. Like using sweets and promises of new toys to negotiate with a stubborn child refusing to leave its hiding place, they offered other elements as tempting alternatives. They offered carbon to lure the oxygen away – the normal approach to refining metals such as iron and

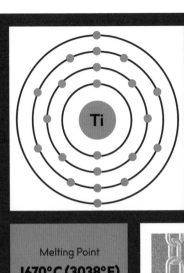

Melting Point
1670°C (3038°F)

Boiling Point
3287°C (5949°F)

Group
4

Period
4

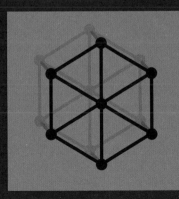

Titanium
Transition Metal

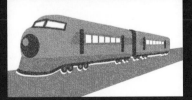

copper. The titanium released the oxygen but grabbed the carbon to make titanium carbide, an even more intractable material. The element held out until 1910, by which time a sufficiently complicated and expensive inducement had been devised. It was worth the effort.

Titanium is exceptionally strong for its weight, and its affinity for oxygen means the pure metal quickly forms a protective oxide layer when exposed to the air, meaning that chemical and physical attacks are shrugged off. It is the perfect material for aerospace, architecture and personal adornment. Titanium and its alloys make lightweight but strong plane parts that will not corrode. The sheets of titanium covering the Guggenheim Museum in Bilbao have the same metallic sheen they had when they were put in place in 1997. Jewellery made of titanium is comfortably lightweight and the oxide layer can be manipulated to different thicknesses that diffract light into a rainbow of colours.

Humans have borrowed titanium's coat of armour in order to benefit from its extraordinary powers of indifference. Pure titanium oxide deflects all sorts of things, including light. Brilliant white titanium oxide makes long-lasting paint. It can also be used to enhance our own protective coating, the skin. Added to sunscreen, it reflects away the UV light that can cause skin cancer.

But while titanium oxide is as indifferent to skin as it is to anything else, the way in which the body reacts to titanium metal tells a different story. Biology can be brutal. The need to defend biological systems from harmful invaders means immune systems are constantly on the warpath, looking to destroy anything unfamiliar using any chemical or physical method available to them. The variety and adaptability of living things means every chink in every armour can be probed for vulnerability. So, the discovery that living systems did not just ignore titanium, but treated it like one of its own, came as a shock.

In 1952, Swedish doctor Per-Ingvar Brånemark wanted to know how new blood cells were made inside the body. To see what was going on, he made holes in the femurs of rabbits but needed a window to cover them. He produced sheets of titanium so thin they became transparent to strong light. With his observations satisfactorily concluded, Brånemark tried to remove the expensive titanium screens for more experiments, but they would not budge. He made new windows, but the same thing happened. Osteoblasts, the cells that form new bone, were attaching to the titanium as if it was bone. Titanium was fully accepted into the body, revolutionizing the field of prosthetics.

Despite appearances, titanium is not perfect. It takes considerable effort, but once provoked, this element reacts violently, and its superhero abilities make it a terrifying opponent. If titanium starts to burn it cannot be stopped. Fortunately, unlike the gods it was named after, titanium takes little heed of the abuse hurled at it, preferring to remain behind its almost impenetrable mask.

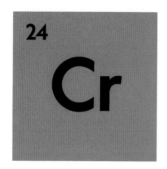

24
Cr

Chromium
The Fading Star

Chromium's problem is that it is easily influenced. Left to its own devices it is the model of clean living, but when it mixes with the wrong crowd the results can be ugly. This is not necessarily chromium's fault; it is just that some other elements take advantage.

The twenty-fourth element in the periodic table sits roughly in the middle of the top row of the long, low rectangle at the centre. The elements in this four-storey block are known as the transition metals. They have all the attributes you might expect of a metal – shiny and solid (mostly) – but they are also slightly flashy.

Chromium, in its pure form, is the silvery metal that gave the high-end finish and shine to state-of-the-art 1920s design. It leant its gleaming, smooth surface to the cars and kitchen appliances so beloved of 1950s America. Its highly reflective finish gave an air of modernity and sleek style that has now faded into a pleasing nostalgic feel.

Transition metals do not just look good in their pure unadulterated state. In the company of other elements, they can produce a dazzling array of colourful compounds, and it is all down to their generosity. Should another atom be in need of an electron or several, the transition metals are usually willing to give up some of their own. They will also

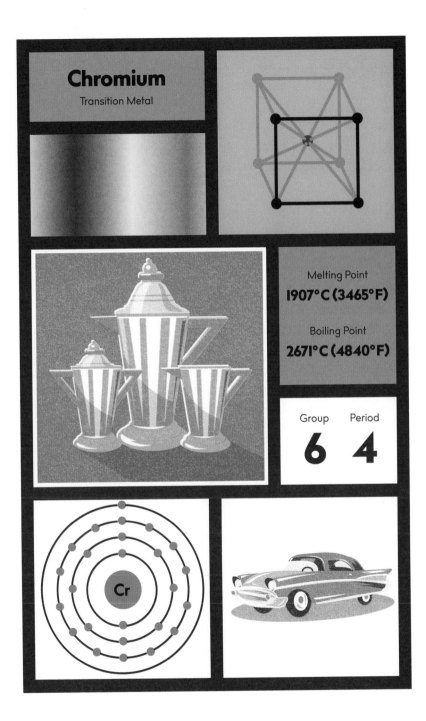

Chromium

Transition Metal

Melting Point
1907°C (3465°F)

Boiling Point
2671°C (4840°F)

Group **6**

Period **4**

Cr

hold on to a few extras should another atom be temporarily over-burdened. They are nothing if not versatile.

Chromium will cheerfully give away or take on extra electrons, but it is most comfortable losing two, three or six, forming chromium(II), chromium(III) or chromium(VI) compounds. Thus, a stunning range of bright-yellow, orange, violet or green compounds can be made just by taking away the right number of electrons and arranging other atoms in the right conformation around chromium. It is why these compounds have been used by artists as pigments.

Many nineteenth- and early-twentieth-century artists fell for the brilliant colours of chromium paints. Vincent van Gogh was captivated by the colour yellow and there were several yellow pigments he could choose from. Cadmium yellow was bright and opaque, but it was expensive. Chromium yellow had a similar brilliant hue, and it was cheaper, but the colour did not last.

On its own chromium can resist tarnishing, but too much socialising with other elements can take its toll. Chromium yellow is known as a 'fugitive' pigment because its brilliant hue slowly disappears behind a dull sheen. The artists that fell for the brilliant colours of chromium paints knew that they had to make allowances for their changeability.

'Paintings fade like flowers,' van Gogh wrote to his brother Theo. When he arranged sunflowers in a vase, he must have topped up the water to keep them alive for a few more days while he painted. And he topped up the paint on his canvases to keep the yellows alive for a few more years. The problem was not just the pigment he used; it was the oil that made the paint flow. Oil paints dry very slowly, so drying agents are added to speed things up. These agents work with oxygen from the air to encourage the chains of oil molecules in the oil paint to join up and form an interlinked mesh that keeps everything in place. Oxygen and drying agent are trapped inside the net that is knitted together around them.

The drying agents van Gogh used also contained transition metals (iron, cobalt or manganese) with similar properties to the transition metals in the pigments – a relaxed attitude to electrons. The combination of oxygen, drying agents and oil released electrons from one transition metal that could then be mopped up by another, and bright yellow chromium(VI) would become dull brown chromium(III).

The study of van Gogh's paintings has taught the art world about how these pigments do, but more often do not, stand up to the passage of time. The sunflower paintings that draw crowds to museums today are a muddy shadow of what they once were.

Until conservators can work out a way to reverse the damage, the best that can be done is to slow the relentless darkening. Blue-green light can encourage metals trapped in the paint to release their electrons which then get soaked up by the ever-accommodating chromium. Changing the lighting in galleries may be the best hope for preserving van Gogh's fading flowers.

26

Fe

Iron
Hard as Nails

Some elements are so well known, their names have bled into our everyday language. Iron is a byword for strength and inflexibility. We talk of an 'iron will', a 'cast-iron alibi' and an 'iron grip'. It's a lot for an element to live up to, especially when that element has a serious weakness, an addiction that can completely undermine its character. In the presence of oxygen, iron just falls apart.

Iron embodies what we have come to expect of all metals. It is heavy, strong, shiny and useful. And, as long as it is kept away from its destructive relationship with oxygen, it more than lives up to those expectations. But, under oxygen's influence, iron becomes brittle and difficult to work with. Iron oxide is the rust that can slowly eat away at the metal until it consumes it completely. It's surprising that a metal so prone to flaky, brittle behaviour would be so widely used.

One major advantage iron has is its abundance. There is no shortage of iron on this planet, although most of it is locked up with its beloved oxygen. The bonds they form aren't especially strong and nature has found numerous ways to separate the two and put the metal to use. Humans took a little longer to figure it out. However, once they established the art of burning off the oxygen locked up within iron oxide

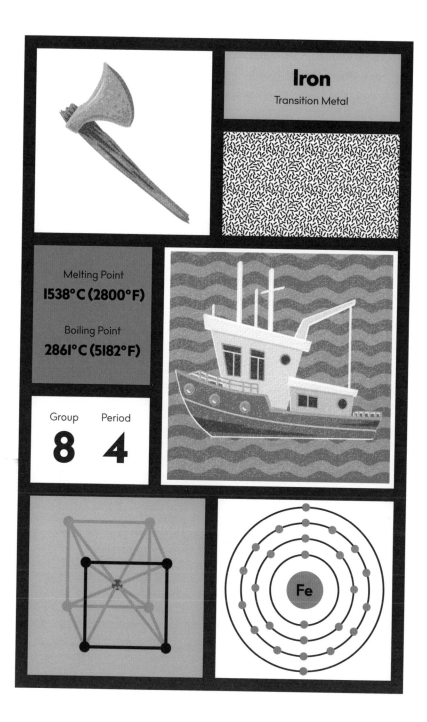

Iron
Transition Metal

Melting Point
1538°C (2800°F)

Boiling Point
2861°C (5182°F)

Group **8**   Period **4**

Fe

ores, there was no looking back. The metal was stronger and kept its sharpness better than any other metal used before it. Stronger tools and weapons gave a competitive edge to those who had unlocked the secrets of iron technology. With time, the process was refined and scaled up to industrial proportions. The scale of the objects made from this metal also grew – from the personal to items that carried persons: boats, bridges and trains. Away from oxygen, iron can achieve incredible feats of engineering, but…there is always a but.

Iron is an incurable recidivist. It has to be constantly looked after and watched over to make sure it isn't making sneaky assignations with oxygen. Water, the key collaborator in iron's downfall, must be excluded at all costs. Physical barriers – paint, grease and plastic – have to be put in place. Iron can also be protected by other metals that act as honey traps for fickle oxygen. Galvanized and stainless steel diverts oxygen's attentions away from iron and towards its preferred zinc and chromium.

Humans go to great lengths to keep oxygen away from the iron objects they create, but inside our body it is a different story. We carry about four grams of iron around with us in our watery bodies. Most of it is in our red blood cells where its express purpose is to interact with oxygen. Day after day, minute after minute, we allow the iron in our blood to come into direct contact with oxygen, and yet we don't rust.

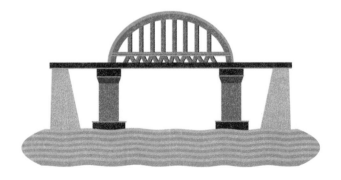

Nature has found a way to use iron's obsession with oxygen to its advantage. Iron has no problem grabbing hold of oxygen, it is excellent at scavenging it out of the air we breathe. It will also hold on tightly to it as it flows through our arteries and capillaries, into every nook and cranny. The trick is persuading it to let go.

The interaction between iron and oxygen in our red blood cells is carefully orchestrated by haemoglobin. Every molecule of haemoglobin contains four haem units. Haem is an elaborate web of carbon and nitrogen atoms that acts as a support network for the single iron atom held at its centre. When the first oxygen molecule binds to an iron atom, the surrounding structure flexes to accommodate its cargo. The movement of one haem is transmitted to the others through a system of molecular springs and levers that ease oxygen binding to the next haem, and the next. Haem allows the oxygen and the iron to form a bond strong enough to stay together until they reach the most distant parts of the body. But, this supportive framework is also there to ease their inevitable separation. When the time is right, this cooperative system gently pushes the oxygen away, but keeps iron wrapped up and ready to go round again.

Iron isn't perfect. Other metals are stronger, more resilient and easier to work with. But, in thousands of different ways, iron is preferred, flaws and all.

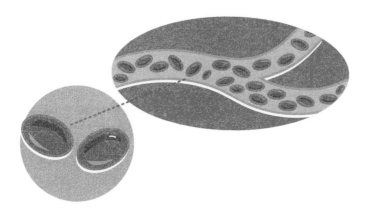

Cobalt
The Trickster

A long time ago, in a land not so far away, humans shared their world with fairies and spirits and goblins. However, these creatures did not always coexist companionably. Medieval miners not only had a physically demanding job, but they had to work in dark, damp mines and deal with the pranks played on them by the goblins that lived there. The miners avoided the tunnels where they could hear the goblins digging. But they were not to be so easily ignored.

One of the goblins' favourite tricks was to disguise useless minerals as something more valuable. They would fool miners into thinking they had found silver buried deep within the rocks. The men would work hard, digging out what they thought were precious silver ores and dragging them to the surface. They would roast the minerals to get at the silver but all they found was worthless metal. Worse still, the fires released choking, poisonous fumes. They cursed the ore and the goblins (or *Kobolde* as they were known in Germany) who had put it there.

In the 1730s, a Swedish scientist named Georg Brandt searched within samples of goblin ore. He found no supernatural creatures, but a previously unknown element instead. He named the element cobalt. Brandt had tried to sort science from superstition, but some of the

Cobalt

Transition Metal

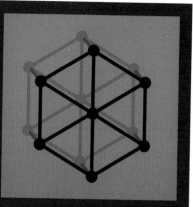

Melting Point
1495°C (2723°F)

Boiling Point
2927°C (5301°F)

Group **9**

Period **4**

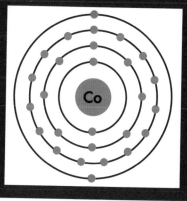

malign influence remained. Cursed, like Cassandra of Greek myth, Brandt told the truth about this new metal, but not everyone believed him. His discovery was not confirmed until after his death in 1768.

The ore the medieval miners had dug out certainly wasn't what it had seemed. It contained no silver, that was true, but the metal wasn't worthless, and it wasn't the cobalt that had been poisoning them. The miners had extracted smaltite, a mineral of cobalt and arsenic. On heating it, they had been poisoned by toxic arsenic fumes released, not the cobalt. The metal also had value. It could be used to make brightly coloured pigments that were added to glass, paints and pottery.

That cobalt could be used to make blue pigments was not a new discovery. The ancient Mesopotamians had blue glass made that colour by cobalt compounds. The Chinese used cobalt to glaze pottery, as did Persians, and, eventually, Europeans. Black/olive grey cobalt oxide was painted onto pottery and porcelain. In the fierce fires of the kiln, it transformed into a bright blue colour that didn't fade over time.

Around the eighth or ninth century CE, the Chinese refined their techniques, heating cobalt ores with alumina to produce cobalt blue. Even so, the cobalt within the pigment remained hidden and was unknown as a unique element. A thousand years later, Europeans figured out how to make cobalt blue for themselves and went into mass production. It soon filled pottery workshops and paint pots. With some tinkering, cobalt could also be modified to produce green, violet and yellow colours. And, while many of these cobalt pigments were appreciated for their permanence, the element could still play a trick or two.

In the seventeenth century covert messages were written in invisible ink. The secret of the ink's composition was only revealed in 1700 by French chemist Jean Hellot. By dissolving cobalt in aqua regia, pale pink cobalt chloride crystals were formed. These crystals could be mixed with water and a little glycerine, to make an ink that was all but invisible when it dried on the paper. A second innocuous message could then be written over the top at right angles to disguise the secret scribblings underneath. Heating the message drove the water from the cobalt chloride crystals, turning them dark blue.

The same trick was used to entertain Victorians with weather-forecasting artificial flowers. The petals were painted with cobalt chloride. They would appear pale pink in wet weather and violet when it turned warm and dry. The colour deepened towards blue as the temperatures rose and humidity fell.

But it isn't always fun and games. Pure cobalt metal is hard wearing, resistant to corrosion and slightly magnetic, useful talents in many applications, good and bad. During the Second World War cobalt was found in another kind of mine. Anchored offshore, these explosive mines were triggered by magnetic fields generated by the ships that floated above them. Cobalt also has important roles in biology, but it can play havoc with our health if there is too little or too much. We always need to be wary of cobalt's trickster nature.

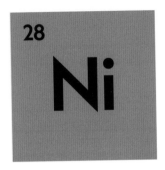

Nickel
The Alien

It all started when millions of meteorites rained down on Earth. It would have been a spectacular sight had anything been around to see it. These arrivals from outer space buried themselves in the ground – near the surface – just waiting to be found. Fortunately, there was no triffid-style uprising when humans did finally stumble across them.

The ancient people who first examined these alien objects must have been surprised at their strangeness. These dark, heavy lumps were like nothing else they had seen and had capabilities far in advance of anything contemporary technology had to offer. The invaders were made of an unusual metal that was incredibly hard and impressively resistant to corrosion. Ancient people used it to make weapons and special items that were revered by those that possessed them. What they didn't realize, was that much of this unique metal was made up of ordinary iron. The superior qualities were due to another, unidentified element.

Around 200 BCE the Chinese produced an alloy they called *pai-t'ung*, or white copper, though it contained no copper at all and was in fact hosting this alien presence. The alloy was exported to the Middle East from where it may have reached Europe, even though the extraterrestrial element was already there, as well. In Europe, too,

Nickel

Transition Metal

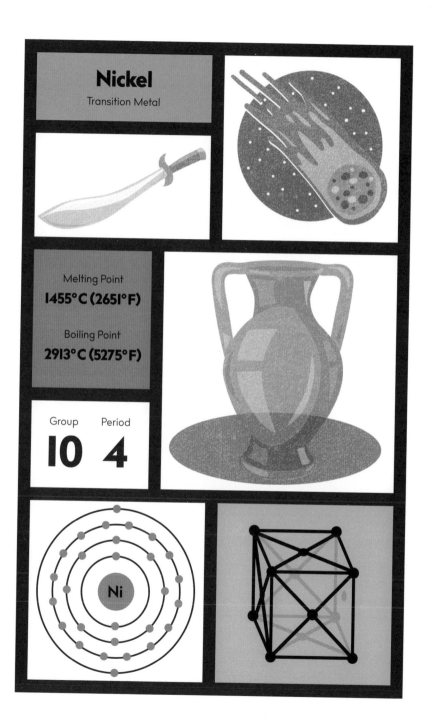

Melting Point
1455°C (2651°F)

Boiling Point
2913°C (5275°F)

Group 10   Period 4

Ni

medieval miners initially mistook the invader for copper ore but, no matter what they did, they couldn't find any copper within it. What they did manage to extract had no apparent use except for colouring glass green. The miners named it Kupfernickel (devil's copper), after Old Nick, because it was the very devil of a mineral.

When the Swedish chemist Axel Fredrik Cronstedt investigated the same mineral in 1751, he also failed to extract copper. Instead he realized that the metal lurking inside the ore was something unique, something that had previously gone unnoticed. He named it nickel. No one believed Cronstedt, thinking his new element was an alloy of several other already known metals. Swedish chemist Torbern Bergman put an end to the argument in 1775 by producing a lump of pure nickel. Finally, the alien invader had revealed itself.

Though its identity had been established, nickel was still largely ignored, mainly because no one could find a use for it. Nickel's advanced abilities remained unknown until 1844 when silver-plating took off. This strange metal proved to be the ideal base on which to coat a layer of silver. And, the more people investigated, the more they appreciated nickel's technological superiority. It was added to coins, cutlery, clocks and thousands of other everyday objects that needed to behave in a reliable and consistent fashion.

The alien slowly infiltrated human lives in dozens of different ways. It has caused us little harm and, on the contrary, has been of incalculable help in developing our technological capabilities. The element has so many everyday applications that it has become almost conventional. H. G. Wells would have been disappointed with these benevolent invaders, but then he didn't live to witness the metal's most mind-boggling ability.

In the late 1950s American metallurgist William J. Buehler was playing around with one-to-one mixtures of titanium and nickel. He named this new alloy Nitinol, after its two components and his work-place (Nickel, Titanium, Naval Ordinance Laboratory). He folded and shaped some samples of the metal into concertinas, to take to a man-agement meeting. The pieces were passed round the table and bent and twisted into new shapes to see if they would break. The associate tech-nical director, David S. Muzzey, idly heated one of the samples with his pipe lighter. To everyone's amazement, the metal moved of its own accord. It flexed and twisted back into its original concertina shape.

Despite appearances, this was not an alien invader suddenly awak-ening from deep stasis. Unlike most alloys that have their component elements randomly placed throughout their structure, Nitinol's nickel and titanium atoms are in a regular alternating arrangement. When Buehler's samples were bent, the atoms shifted slightly, like dominoes lying flat. Heat restored them to their original positions and the object returned to its original shape.

Nickel is not an entirely alien metal, though much of what is mined for our daily use is probably extraterrestrial in origin. The element has been part of our planet all along, mostly hidden away in Earth's core. Combined with iron, its magnetic properties produce the protective planet-sized shield that deflects the dangerous charged particles that continuously bombard our planet. Nickel's strange properties are both alien and very close to home.

29
Cu

Copper
The Blue-blooded One

If you were to cast copper in a role, it would be the noble knight errant. This element has a long and distinguished heritage, and blue blood flows in its veins. Copper isn't quite top tier, like gold or platinum (elemental royalty), but it is certainly a cut above common metals. It wears its pinky-orange finery like glittering armour. Its beautifully coloured ores are like the banners and elaborate heraldic shields of old. And, in true chivalric style, it battles against the bad guys.

As with any aristocrat, copper has an excellent pedigree. Its untarnished line stretches back through ten thousand years of human history. It is one of the few metals that can be dug out of the ground in its pure form. And, though it does combine with other elements, it naturally gravitates towards socially acceptable marriages. Copper ores such as malachite and azurite (variations on copper carbonates) are prized for their rich colours and ornate patterns. They have been polished into decorative objects and ground into pigments to adorn the faces of the ancient rich and powerful.

A rich appearance is important, but nobility has little to do with money. Like many aristocratic fortunes, copper's value has undergone many peaks and troughs over the centuries. Though copper is used in

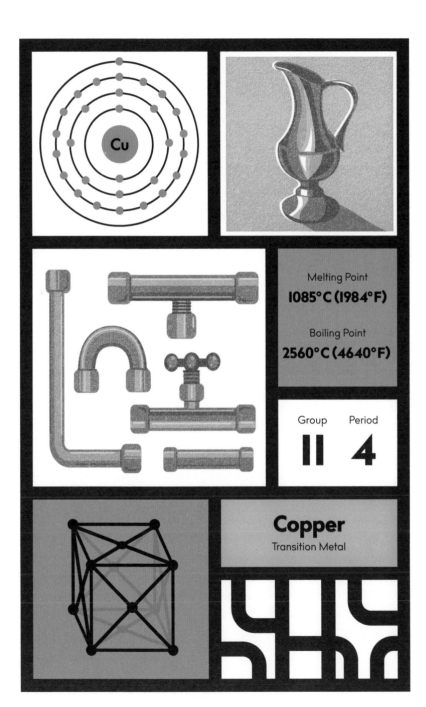

Melting Point
1085°C (1984°F)

Boiling Point
2560°C (4640°F)

Group
11

Period
4

Copper
Transition Metal

coins, few countries have based their wealth on copper reserves. One exception is Sweden where the Great Copper Mountain at Falun supplied so much metal over the nearly one thousand years it was in commercial operation, it funded several wars. In the eighteenth century this one mine satisfied one-third of all Europe's copper demands.

In addition to make-up products that enhanced royal faces in the past, copper also gives emperors' blood its blue colour. The emperor in question is the emperor scorpion (*Pandinus imperator*), who, along with other arthropods and cephalopods, have blue blood from the copper they use to carry oxygen. Pairs of copper atoms are incorporated into the enzyme haemocyanin, so named for the blue colour it turns when it binds with oxygen. But, the emperor scorpion did not get its regal title because of its blue blood, it had more to do with its impressive figure. It is the largest of all scorpions and sports head-to-stinger black armour that shines with a metallic green lustre.

All humans have the base metal iron at the centre of their oxygen-carrying system, giving our blood its distinctive red colour; princes and paupers alike. Iron-centred haemoglobin is more efficient than the copper-based haemocyanin flowing through the emperor scorpion's body, but it works less well at lower temperatures, or where there is less oxygen available. The human nobles that claimed they were blue blooded were lying. They took advantage of their pale complexions to show up their blue veins just below the surface of their skin. A trick of the light filtering through skin and fat that makes the vein appear blue.

We humans may not have blue blood, but copper still helps us breathe. Enzymes in our cells that help us process oxygen to release energy are copper dependent. And there are many other ways copper assists us. It shelters us from the rain in roofing material. On the bottoms of ships, it kept sailors afloat by deterring ship barnacles that could compromise the wood underneath. To be copper-bottomed is to be reliable or trustworthy, traits any knight errant would be proud of. Furthermore, in true knightly fashion, copper also takes up the fight against our oldest enemies – bacteria. When bacteria settle on the surface of copper they are killed. Using copper in coins, and lining air-conditioning units with it, helps vanquish our foe.

But, like one of the most famous knights errant, Don Quixote, who loses his mind after reading too many chivalric romances, too much copper can be a bad thing. An excess of copper may not result in tilting at windmills but it can do serious damage to the body.

Don Quixote parodied the notion of a Golden Age of Chivalry. The same romantic ideal depicted in pre-Raphaelite paintings, of knights wooing copper-haired maidens, never really existed. Today, knights won't win the hearts of fair maidens by jousting or embarking on quests for the Holy Grail. But copper has certainly not fallen from favour. Old-fashioned qualities, such as purity and resistance to taint, made copper a good choice for plumbing. Newly discovered traits, such as its excellent ability to conduct electricity, means demand for the metal is now higher than ever.

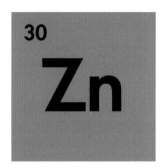

30

Zn

Zinc
The Wallflower

Zinc's presence passes most people by. As one of the transition metals, an exuberant, charismatic family of elements, you would expect it to stand out more. But, even in a family of extroverts, there are those who fail to shine as brightly, or shout as loudly. These individuals are always there, but somehow manage to fade into the background even when they are standing right in front of you.

Tucked away in the far-right-hand corner of the transition metal family tree, zinc has none of its relatives' flare. Compared to its cousins, gold, copper and mercury, the appearance of pure zinc is rather uninteresting. Its fellow transition metals are known for their colourful compounds, but zinc sticks to a more limited palette ranging from colourless to white. The configuration of zinc's electrons makes it more conservative than its brethren. That doesn't mean zinc hasn't found an incredible number of uses, it just doesn't make a song and dance about it. Such undemonstrative behaviour means it is easy to overlook.

Zinc was officially 'discovered' in 1746 by German chemist Andreas Sigismund Marggraf, despite the fact that humans had already been mining and refining it for at least 600 years, and using its ores for thousands of years before that. Objects have been found, fashioned

Melting Point
420°C (787°F)

Boiling Point
907°C (1665°F)

Group
12

Period
4

Zinc
Transition Metal

from 90 per cent zinc, dating from 2500 BCE. Bronze Age people processed zinc and copper ores to produce brass. Strabo, a Greek philosopher writing in the first century BCE, talked about 'mock silver' as a metal distinct from brass. And yet, the metal seemingly failed to hold anyone's attention long enough for anyone to give it its own name or bother to recognize it as a separate and unique entity.

Misidentification, or failing to understand the mixture of metals within alloys is one thing, but not noticing four centuries of zinc production that churned out an estimated 1 million tonnes of the metal, is quite another. Zinc was recognized as a metal in its own right in India around the fourteenth century. Europeans, meanwhile, managed to remain oblivious to what was right in front of them for another 300 years. Shipments of zinc arrived from India, and alchemists burned the metal to produce zinc oxide, what they charmingly called 'philosopher's wool'. In 1668, P. M. de Raspoour, a German metallurgist, isolated the metal. A few decades later Geoffroy the Elder did the same and wrote about it as something new. By 1743 a zinc smelter was operating in Bristol, England, producing two hundred tonnes a year. But these exciting discoveries, or rediscoveries, failed to make any impact.

Marggraf's main achievement in 1746 was to give zinc the recognition that was so long overdue. Then, once people started to look at zinc

for what it was, they realized just how blind they had been. It might be a very unprepossessing element, but zinc is incredibly useful.

Zinc metal is easy to work and resists weathering. It can be used to protect buckets, boats and buildings. Paris's distinctive architecture, recorded by so many Impressionist painters, is capped by zinc roofs. The paints the Impressionists used contained zinc oxide, a brilliant white pigment that is also included in cosmetics, paper and sunscreen. With other metals, zinc can make brass, galvanized steel and a whole host of useful alloys. Paired with copper, zinc was used to make the first batteries over two hundred years ago; it's still used to make them today.

The list of zinc's achievements goes on and on, and does so with little fuss. There are no major environmental worries over zinc. Its widespread use in industries raises relatively few concerns. Personal and domestic applications should not cause too many sleepless nights. And all because zinc has had even more success in biological systems.

Humans were so oblivious to zinc that it wasn't until 1869 that anyone suspected it might be essential for health, at least for some fungi. In the 1920s, scientists speculated whether the zinc they found in human tissue had simply been absorbed from the environment, or was actually performing some function. We now know that zinc is not just essential to life, it is ever-present. Humans alone contain thousands of zinc-dependent enzymes that help us digest, reproduce, respire and think. Sometimes it is at the heart of the actions being carried out by an enzyme, other times it is purely structural, but it is always important. Zinc is the embodiment of the understatement.

31

Ga

Gallium
The Francophile

Gallium's coming was foretold, and by none other than the chemical prophet Dmitri Mendeleev. It was certainly an important moment in science, but the addition of gallium to the periodic family tree was not received with the glorious adulation and unstinting praise some may have thought it deserved. Scientific egos were left bruised by public spats and practical jokes. Gallium, by and large, managed to rise above it all. It also had a few tricks up its sleeve that astonished everyone.

When Mendeleev drafted his periodic table, he noticed gaps; missing elements he was sure existed but that hadn't been discovered yet. He published his table with annotations as to what these missing elements would be like and he gave clues as to where they might be found. Element 31 should be a lot like element 13, aluminium, because they are both members of the same chemical family. It might get mixed up in aluminium ores and, sure enough, that is where a lot of gallium is to be found.

In 1875, Paul-Émile Lecoq de Boisbaudran was examining a zinc ore when he spotted signs of another element locked up within its crystals. He managed to isolate a metal from the mixture, one that he suspected no one else had seen before. He claimed the discovery of a new element

Gallium

Triel

Melting Point
30°C (86°F)

Boiling Point
2229°C (4044°F)

Group
13

Period
4

and, in a fit of nationalistic pride, he named it gallium, after Gallia, the Latin name of his native France.

Not everyone was convinced of Boisbaudran's patriotism. Some thought his pride was rather more self-centred and accused him of deriving gallium not from Gallia, but from *gallus*, the Latin for cockerel so seeking to honour himself, Lecoq (French for cockerel), rather than his country. Such blatant egotism could not be tolerated by the scientific community and Lecoq was obliged to deny he had ever even considered the association with his own name. It wasn't his only problem.

Lecoq was unaware of Mendeleev's table when he made his discovery. When the Frenchman reported the discovery of gallium in the literature, Mendeleev tried to claim it as his own, having predicted its existence and so many of its characteristics. Lecoq bluntly pointed out that he had done all the hard work and so deserved the credit. He also dared to suggest that an obscure French chemist, and not the Russian Mendeleev, had been first to design a table of the chemical elements. Mendeleev was having none of it.

The Russian chemist was obviously the original deviser of the periodic table, but he conceded that Lecoq was the discoverer of gallium. However, he was going to prove he knew more about the element than Lecoq, without ever having seen it. After careful study of the published data, Mendeleev noticed the values given for gallium's density and weight were considerably lower than he had predicted. He wrote to Lecoq to tell him his sample wasn't pure and he should refine it.

So, Lecoq went back to his lab. To his undoubted chagrin, analysis of a pure sample of gallium only reinforced Mendeleev's position. The true density and weight did indeed match the Russian's prediction. But Mendeleev hadn't foreseen everything. This element had surprises in store for everyone.

Gallium certainly has chemical characteristics that are similar to those of aluminium, as Mendeleev had predicted. But it also somewhat resembles zinc, as Lecoq found, and can do a passable impression of yet another element, mercury. Gallium's low melting point means it will turn into a silvery liquid in the palm of your hand. Chemists could trick each other by handing over spoons fashioned from gallium that would disappear into a colleague's tea as they stirred it. And gallium remains liquid over a wider range of temperatures than almost every other substance, making it a less toxic substitute for mercury in high temperature thermometers.

Inside the body, gallium could also be mistaken for iron, being taken up by iron-centred enzymes needed for everyday growth and repair. The mistake can lead to disrupted cell regulation and replication. This is not necessarily as devastating as it may first appear. Cancer cells need a lot of these iron-centred enzymes to support their high metabolism and rapid rates of replication. Gallium drugs are therefore concentrated in these cancerous cells, throwing a spanner in the tumour's works. For all his self-assurance, Mendeleev couldn't have known gallium would melt so easily, or that its toxic properties could help heal the body. Lecoq, undeterred by past mistakes, went on to discover two more elements (samarium and dysprosium), but he wisely chose less controversial names for them.

33

As

Arsenic
The Poisoner

Arsenic is the gold standard of chemical killers. In fiction and fact, it has been used to bump off wealthy relatives, rivals and other innocents who stood in the way of someone's ambition. Arsenic is abundant, tasteless and its symptoms of poisoning can be blamed on natural causes. It may seem the perfect weapon, yet its prolific use in the nineteenth century led to the development of a simple, reliable and very sensitive test. It spawned the field of forensic toxicology and thwarted many a potential poisoner.

Thanks to the test, more cases of suspicious death were brought before the courts. The accused also now had to explain just how a lethal dose of arsenic had found its way inside the body of their nearest and wealthiest relative. Some simply stated the unfortunate victim must have eaten it. Then, perhaps with a shrug of the shoulders, they might helpfully add, clearly, they ate too much. This was more than pure bravado. Despite its well-known toxicity many people in the Victorian era did indeed choose to eat arsenic.

It might seem odd that an element commonly used as rat poison would be considered safe for consumption. Victorians may have been relying on the advice put forward by the eminent natural philosopher

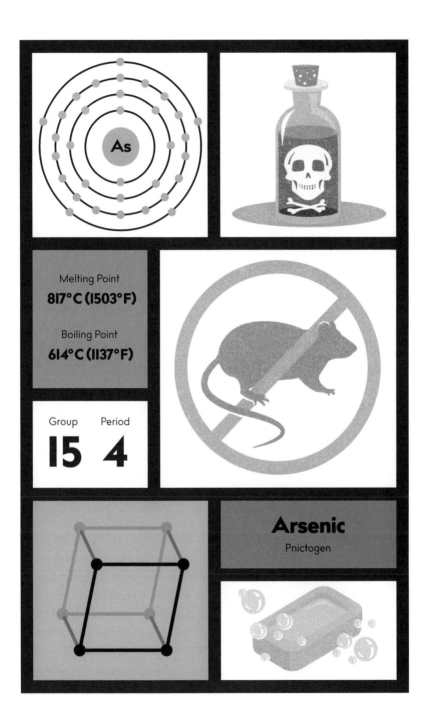

As

Melting Point
817°C (1503°F)

Boiling Point
614°C (1137°F)

Group
15

Period
4

Arsenic
Pnictogen

and 'father of toxicology', Paracelsus, who in the sixteenth century stated that it was 'the dose that made the poison'. He wasn't wrong, but Victorians were particularly cavalier in their interpretation.

The arsenic-eating phenomenon started in Styria, a mountainous region of Austria. People working at high altitude or in deep mines doing hard physical labour found that eating tiny amounts of arsenic helped them breathe more easily. A small lump, about the size of a grain of rice, was crushed between the teeth or grated onto toast two or three times a week. These men also found the arsenic bulked them up a little bit, making them look more muscular, and more attractive. Before long it was being taken as a beauty aid. Women started to swallow it because it gave them gorgeous curves. Everyone, men and women, benefited from perfect, flawless complexions and thick, glossy hair. They all appeared to be in rugged health and the practice soon spread across Europe.

Manufacturers were keen to cash in on the craze and started adding arsenic to soaps and face washes, as well as dissolving it into tonics and pick-me-ups. Those who could not afford branded arsenic beauty products made their own by soaking fly-papers and dissolving rat poison. It may have made them look great, but it was all an illusion. The clear complexion was from the arsenic killing the bugs and bacteria that would normally cause spots and blemishes. The bulking up and beautiful curves were oedema, the accumulation of fluid in the body, and an early sign of ill health. The glossy hair was the result of arsenic bonding tightly to its favourite element, sulphur, found in abundance in hair and skin.

It was arsenic's love of sulphur that caused all the problems. There is a lot of sulphur in our bodies, meaning arsenic is easily trapped there. Over time, even small doses will accumulate inside the body causing all manner of problems. Sulphur atoms in proteins and enzymes hold these molecules in a certain shape, like molecular Velcro, so they can carry out their jobs. With arsenic in the way, the enzymes do not work so well. Symptoms vary as different enzymes are affected, but the accumulation of arsenic eventually leads to loss of vital functions and death.

The arsenic-eaters of Styria could not maintain their good looks forever. Eventually their pale, peaches-and-cream complexions would become blemished with dark spots and rough, scaly skin. Even if they managed to avoid the more visible ravages of chronic arsenic poisoning, their time would eventually be up.

Styria had limited space for burials, so internment was only a temporary situation. Years after death a grave would be revisited, and the bones removed to a charnel house to leave a vacant lot. A lot of arsenic in a body will kill off the organisms normally involved in decomposition. Stories emerged from central Europe of perfectly preserved corpses being found in their graves. Adding yet another horror to its long list of terrors, arsenic may have contributed to these tales of the undead.

Se

Selenium
The Stinker

We all have those friends who we absolutely adore, but their forceful personalities mean we can only cope with them in small doses. Too much time in their company, and we begin to wonder why we ever became friends, but we miss them when we don't see them in a while. Selenium is one such friend; a little goes a long way.

Selenium is closely related to sulphur and, as such, easily slips into the same chemical shoes. The amino acids that incorporate sulphur atoms (cysteine and methionine), and are essential to our health, can also accommodate a selenium atom in the same spot, but they are not used in quite the same way. Selenium-substituted amino acids are built into some very important enzymes. One, deiodinase, promotes hormone production in the thyroid gland. Another, glutathione peroxidase, guards us against a small but significant and ever-present threat. Tiny amounts of hydrogen peroxide occur naturally in water. Inside the body, these peroxide molecules can form dangerously reactive oxygen species. Selenium-based peroxidase breaks up peroxides before they can cause trouble.

The importance of selenium is best illustrated by what happens when there isn't enough of it. A lack of selenium has been linked to problems

Selenium

Chalcogen

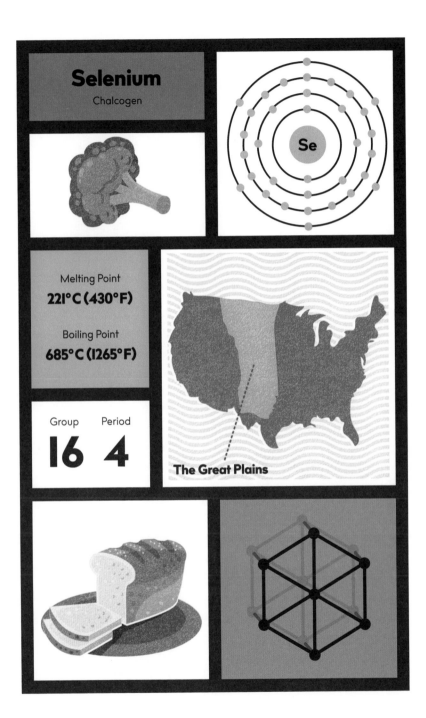

Se

Melting Point
221°C (430°F)

Boiling Point
685°C (1265°F)

Group
16

Period
4

The Great Plains

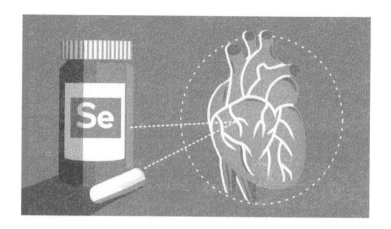

with fertility. Correlations have also been found between selenium levels in the blood and the body's immune response. Keshan disease results in a weakening of the heart. Kashin-Beck disease causes a deforming arthritis. Both conditions are associated with selenium deficiency, which weakens the immune system and allows pathogens easier access to the body. The outcome of infections from the human immunodeficiency virus (HIV) can be predicted by how much selenium is in the blood. A drop in selenium levels and the prognosis is poor.

Eating foods that are rich in selenium will certainly help to keep your blood levels topped up and selenium-based enzymes at an optimum. We get our daily selenium dose in bread, broccoli, brazil nuts and many other foods that don't necessarily begin with the letter b. A more important factor, however, is where the food was grown. Selenium is not evenly distributed around the world. The soil in the Keshan region of China has exceptionally low levels. The great plains of America are blessed with an abundance of selenium. Much of Europe falls somewhere in between.

Such an unreliable supply would seem to be an excellent recommendation for selenium supplements. In some regions where selenium is lacking it is added to animal feed and fertilizers. You may also be tempted to top up your personal intake. But, before you rush to swallow the selenium stocks of your nearest health food supplier, there are some things you should know.

Selenium may be very similar to sulphur but there are important differences. While we have somewhere in the region of 140g (5oz) of sulphur in our body, our tolerance for selenium is ten thousand times lower. The maximum recommended daily dose of selenium is 450 micrograms. More than this is too much. Your body will try to get rid of the excess by converting it into volatile selenium compounds to be breathed and sweated out. Your nearest and dearest will be the first to notice the first and most obvious signs of selenium poisoning – bad breath and body odour that no amount of soap or toothpaste can alleviate. And you don't necessarily have to eat it to experience problems.

Jöns Jacob Berzelius discovered selenium in 1817. The process of extraction, and the tests he performed on the element to identify it, exposed him to toxic levels to the extent that his housekeeper was moved to remonstrate with her employer over his appalling breath. She admonished him for eating far too much garlic, a plant full of odorous sulphur compounds.

Sulphur and selenium are members of the same chalcogen chemical family, but they are also linked by their ability to produce an astonishing stench. The eminent neurologist Oliver Sacks, an enthusiastic amateur chemist in his youth, nicknamed them the 'stinkogens' after his unforgettable experiments with some of their compounds. 'If the smell of hydrogen sulfide was bad, that of hydrogen selenide was a hundred times worse – an indescribably horrible, disgusting smell that caused me to choke and tear'.

If the smell isn't warning enough that selenium should only ever be taken in moderation, there is worse. Too much selenium and your hair can start to fall out, nails can slough off and nerves can be damaged. It is a pungent reminder that you can have too much of a good thing.

Bromine
The Sedative

There is something retro about bromine. Maybe it is the colour. Or maybe it is because one familiar story about the element comes from the world wars of the early twentieth century. Bromine certainly had a few glorious decades of popularity before fading into relative obscurity, like an ageing starlet.

The peak of bromine's fame was when it featured in a 1940 Broadway musical. A line in a song from Rodgers and Hart's *Pal Joey* mentions bromo-seltzer, a well-known remedy at the time for sleepless nights and overstretched nerves. The song was 'Bewitched, Bothered and Bewildered', and, though it is about a woman infatuated with a young man, the song title could just as easily apply to the bromine that features in the song's lyrics.

Bromine is bewitching because of its uncharacteristic characteristics. It is one of only two elements that are liquid at ordinary room temperature and pressure (the other being Mercury). It is also strongly coloured, a rarity amongst pure elements. Its rich, orange-brown hue is like an intense sepia tone and perhaps adds to the old-fashioned feel that surrounds the element. But, before the warm, fuzzy glow of nostalgia creeps over you, there is the 'bothered' part of the song title to consider.

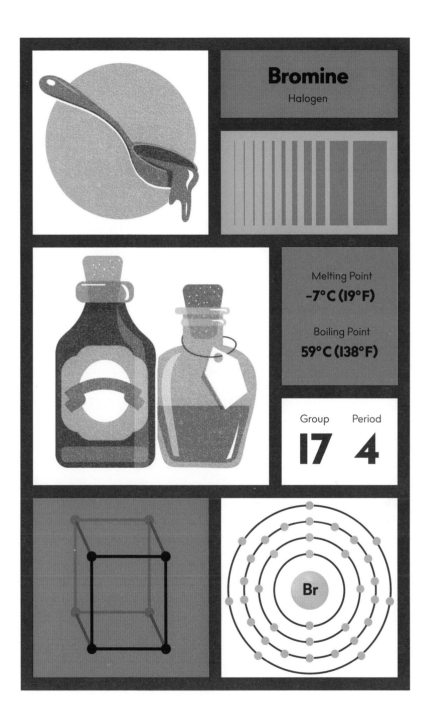

Bromine

Halogen

Melting Point
−7°C (19°F)

Boiling Point
59°C (138°F)

Group
17

Period
4

Br

The halogen family of elements, of which bromine is a member, has a reputation. Fluorine and chlorine are well known for their destructive capabilities. Bromine follows the family trait, though with less enthusiasm than its siblings. Like other halogens, bromine craves an extra electron to complete its outer shell. However, it is bigger and more unwieldy than either fluorine or chlorine, and it simply does not have the same determined pull. While not in the same league, bromine will still have a damaging, corrosive effect on the lungs and throat of anyone unwise enough to sniff or drink it in its pure form. Reacted with a metal to make a bromide salt, the situation is rather different. Content with its full outer shell, bromide is a very different beast, though it still has an appreciable effect on the human body and hence the third part of the song title – 'bewildered'.

A few things about bromine, or bromide, are confusing. For one, how bromide gained its undeserved reputation for suppressing libido. Wherever the belief came from, it inspired Sir Charles Locock, physician to Queen Victoria, to recommend bromide salts in treating epilepsy, which he thought was caused by masturbation. The bromide powders he gave to his epilepsy patients certainly improved their condition, but not for the reasons he suggested. In fact, bromide is a sedative, and it suppresses activity in the brain that can cause epileptic seizures.

Locock's recommendations were made in 1857 and, despite growing evidence of unpleasant side effects, more and more therapeutic uses were found for bromide powders. They were swallowed as both sleeping draughts and hangover cures. By the time Rodgers and Hart penned a song for their musical, bromide powders were being produced in bulk, boxed up and branded for mass consumption as bromo-seltzer.

Relatively large quantities of bromide are needed to produce the desired sedative effects. And, though bromide is easily absorbed into the body, it is reluctant to leave. It can take nine to twelve days to remove just half of what is inside you. When individuals were swallowing several grams on a regular basis, bromide soon built up to toxic levels. Bromism became a problem. The symptoms, lethargy, slurred speech, depression and confusion, were rarely fatal but certainly debilitating. The persistent myth of British soldiers being given bromide in their tea during the world wars seems unlikely to be true. It would be unwise to have servicemen bewitched, bothered and bewildered.

Bromide powders have been almost completely replaced by safer and more effective sedatives. As a result, bromine has rather fallen from the limelight. However, some still hanker after the fashions of the past.

Photos of stars singing about bromo-seltzer in the 1940s were captured by a layer of silver bromide on photographic film. Light reflected from an object onto the film split the bromide from the silver to leave dark metal patches. In dark rooms, the bromide that had been released, together with any unreacted silver bromide, was washed away to reveal a negative image. Although digital photography now dominates, some still prefer the images that bromide salts produce. Covers of 'Bewitched, Bothered and Bewildered' are still being made, and bromine and bromide salts still have roles to play.

37

Rb

Rubidium
The Uninvited Guest

You know those families where every member has a world-class talent? Mum's a brain surgeon, the eldest just made their first million with a start-up company, the middle child is an award-winning choreographer and the youngest is a virtuoso cellist. And, in amongst all the brilliance, is the sibling they never really talk about. The one who is just ordinary, with no particular flare or gift, who muddles through life as best they can, just like the rest of us. Meet rubidium.

Rubidium is the ordinary member of an illustrious family of elements – the alkali metals. It is the one that always gets overlooked and overshadowed by its over-accomplished brethren. Had things been a little different, it could have been a household name like its siblings. Instead, it feels as though rubidium has always been hanging around waiting for an opportunity that never comes and has been relegated to the backwaters of scientific research.

The elements in each vertical group of the periodic table follow distinct trends as you move from one element to the next going up or down the column. From top to bottom the atoms of each element get a little bigger as shells of electrons are added like layers of an onion. The results of these regular, predictable additions are regular, predictable

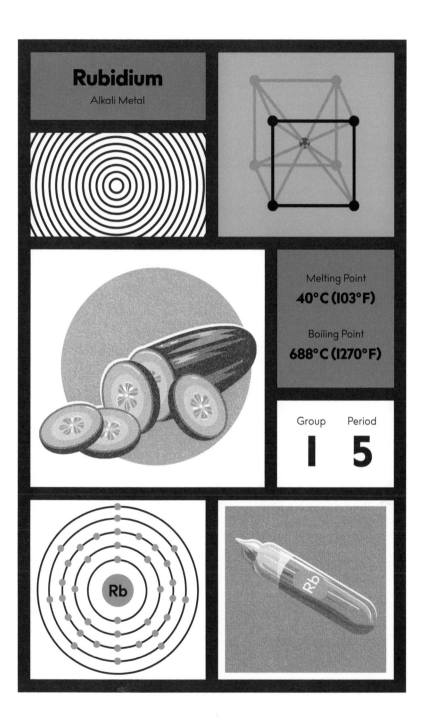

Rubidium

Alkali Metal

Melting Point
40°C (103°F)

Boiling Point
688°C (1270°F)

Group
1

Period
5

Rb

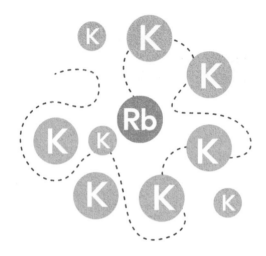

modifications to the properties of the elements. Elements immediately above or below each other can be strikingly similar. Rubidium's similarity to potassium is uncanny. It happily adopts the same single positive charge, it is a very similar size and can do all the same jobs as potassium, almost as well. It would win every potassium look-alike competition, but it is not quite the same thing.

Biological systems are built around the chemical and physical properties of different elements. Sodium and potassium are just the right size and have just the right chemical balance between them to produce the electrical signals in nerves. Because it is so important, the human body is quick to grab hold of potassium in the food we eat. Rubidium can also sneak into our food, and thereby our bodies, because it is also present in our environment and can pass itself off as potassium. You could have half a gram of rubidium floating around inside you right now as a result of mistaken identity. It is the highest concentration of an element in humans that has no known biological role. In fact, rubidium is at a higher concentration inside the human body than in the environment because the body will absorb and hold on to it more readily than potassium. It arrives in our bodies, uninvited but unobtrusive, simply waiting for the opportunity to show what it can really do.

Rubidium is like the uninvited guest at a party who no one seems to know but who mingles in the crowd and gets along with everyone anyway. Once inside the body, it easily slots into potassium's positions and takes over potassium's jobs. It is such a convincing potassium impersonator that you could eat a lot more rubidium and it would almost go unnoticed. Almost. Young rats fed on a potassium-free diet stop growing, but normal growth can be resumed by substituting rubidium – at least for a while. Rats can live perfectly happily if half their potassium is replaced with rubidium but there are limits. When rubidium is fed to rats in large quantities, without some potassium there to help, the slight differences between the two elements become evident. After a few weeks of a rubidium-only diet, the young rats become more and more irritable and eventually die in convulsions as the electrical signals in their nerves misfire.

What is surprising is that the rat's body can tolerate so much of something that should not be there before anything goes wrong. It shows just how close rubidium comes to being on a par with its brilliant sibling potassium. It is only because rubidium is very rare, compared to the very abundant potassium, that our bodies do not contain more of it. Had our planet formed with a higher proportion of rubidium, evolution might have taken a very slightly different path. Our nerves may have produced their electrical signals using sodium and rubidium, and there would have been countless other subtle changes if it had assumed the roles normally performed by potassium. Small mutations lead to huge evolutionary changes, it just takes a lot of time for all the small changes to build up.

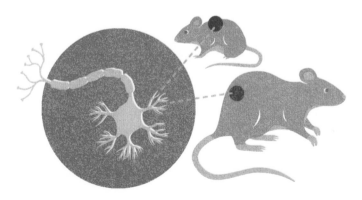

Yttrium
The Russian Doll

There is a tiny village on the Swedish island of Resarö that should be more famous than it is. Ytterby, a quiet unassuming place, has given its name to no fewer than four elements in the periodic table – yttrium, ytterbium, terbium and erbium. Four more elements can trace their discovery back to the same site. No other place on Earth can claim to be so rich in elemental discoveries. And it all started with yttrium.

Just outside Ytterby is a mine rich in mineral deposits. Sixteenth-century miners extracted quartz from its depths, then later feldspar became the commercially important mineral, because of its use in the porcelain industry. But the mine also had a reputation for the wide variety of rocks it contained that yielded unusual pigments and glazes. In 1787 Carl Axel Arrhenius, an artillery officer in the Swedish army, was inspecting the mine when one rock in particular caught his attention. It was black and very heavy. Arrhenius had a keen interest in mineralogy and this rock did not look like anything he had seen before. His chance discovery opened up a scientific rabbit hole. Every time the end was almost in sight, a new twist, blind alley or sidetrack would appear. It took almost a century for scientists to pick apart all the chemical components of the mineral Arrhenius had stumbled across.

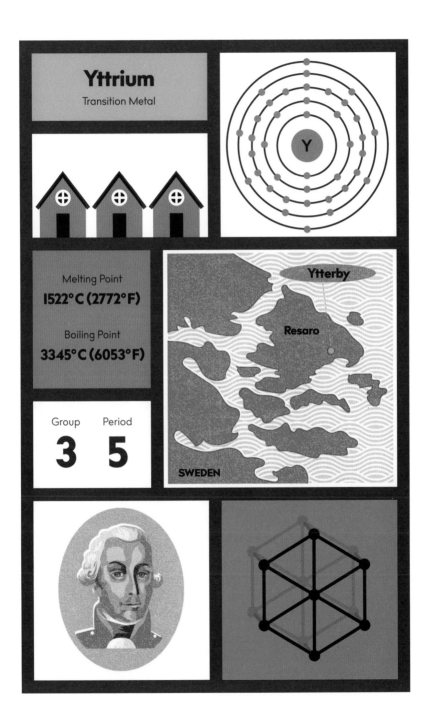

Yttrium

Transition Metal

Melting Point
1522°C (2772°F)

Boiling Point
3345°C (6053°F)

Group
3

Period
5

Ytterby

Resaro

SWEDEN

Thinking he had found a source of tungsten, an element discovered only a few years earlier, Arrhenius sent a sample of the black rock to Johan Gadolin at the university of Åbo for proper analysis. Gadolin found that whatever was giving the rock its great weight, it was not tungsten. Thirty-eight per cent of the stone was made up of a previously unknown substance that he described as an 'earth', an old term meaning a metal oxide compound. Gadolin named this new 'earth' yttria, and therefore the metal part of this metal oxide must be yttrium. He published his findings in 1794 and got the credit for discovering a new element even though he had not managed to separate yttrium from the oxygen to which it was bound. The mineral was named gadolinite in Gadolin's honour, but the work of isolating the yttrium it contained was done by Friedrich Wöhler in 1828. But here comes the first twist in the tale – Wöhler's sample of yttrium was not pure.

In 1843 Carl Gustaf Mosander found that yttria was in fact made up of three different metal oxides – white yttrium oxide, plus a yellow terbia and rose-coloured erbia. These new oxides yielded terbium and erbium metals. Another two elements were added to the growing list, but they were still not at the end of the impurity problem.

A few decades later, erbia was shown to contain a mixture of another three elements, ytterbium, holmium and thulium, while a closer examination of gadolinite samples revealed a seventh element, scandium. Each new metal was surprisingly similar to the last, only the quantities got smaller and smaller. The elements were seemingly stacked inside gadolinite like Russian dolls. Last, and least, was gadolinium, found in 1880 in tiny quantities using spectroscopic analysis.

It was the chemical similarities between these elements that allowed them to clump together in the rocks in the first place. It was also what made them difficult to separate when they were discovered, and the characteristic that came to define the family of elements that they all belonged to. The 'rare earths', as they became known, include seventeen similar-looking metals. They form the long row of fifteen blocks just below the main periodic table, plus two extras, scandium and yttrium, included because they are chemically kindred spirits.

Though they are characterized by their similarities, all the rare earths have distinctive features that have helped them carve out individual niches. Yttrium, for example, combined with aluminium and oxygen, can make garnets that form fake diamonds and the cores of high-powered lasers. It is the sort of technology you might expect a Bond villain to use in their plans to take over the world.

Although few people may have heard of it, yttrium has led to the discovery of a whole new branch of the periodic table and enabled the development of globally important technology, from LED lights to batteries and superconductors to cancer treatments. Yttrium may have had an unpromising start in life, but it has not done too badly for itself.

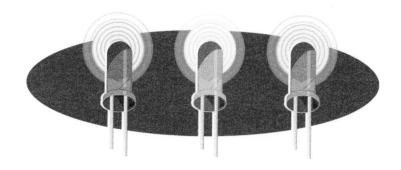

Technetium
The Monster

In 1936 a scientist and his assistant were in their lab searching through scraps of other people's experiments. They were somewhat circumspect about their activities because they had made a tremendous discovery – the secret of making new elements from the fragments of others. The artificial element they unleashed on the world was also very unstable. Had they created a monster?

The scientists were Emilio Segrè and Carlo Perrier, physicists and element hunters. Periodic tables in the early twentieth century contained a few conspicuous gaps. One particularly glaring hole was right in the centre of the table, just next to molybdenum, element 42. Dmitri Mendeleev, father of the periodic table, had predicted the existence of element 43 in the mid-nineteenth century. He had even described the properties it should possess, but no one had been able to track it down.

As the understanding of atoms and elements developed, the prospect of engineering an element became a tantalising reality. Knowing that each element was defined by the number of protons in its nucleus, it was theoretically possible to transmute one element into another by adjusting the number of protons it contained. If element 43 could not be found, perhaps it could be made. The question was how.

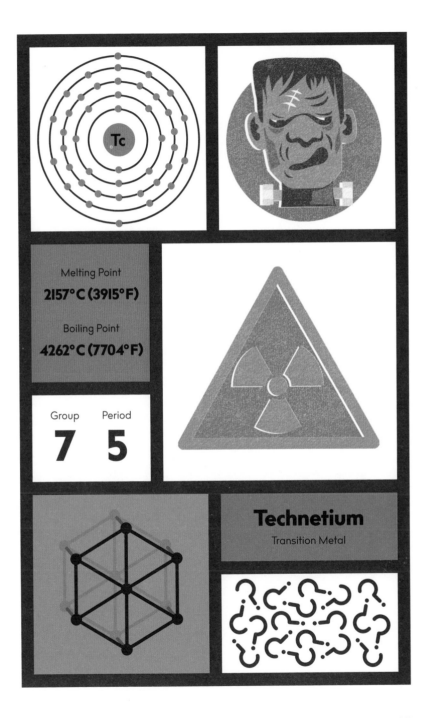

Melting Point
2157°C (3915°F)

Boiling Point
4262°C (7704°F)

Group
7

Period
5

Technetium
Transition Metal

Segrè stumbled across the answer when visiting Ernest Lawrence at the Lawrence Berkeley National Laboratory in California. Lawrence was showing off a new piece of kit, a cyclotron, or atom smasher, to his guest. The powerful machine fired light atoms at heavy atoms in the hope that they would stick together. Lawrence was not interested in making new elements – he had other plans for his cyclotron – but Segrè realized the possibilities. He became particularly excited when he learned that some of the replaceable parts of Lawrence's machine were made of molybdenum. What if a stray proton, debris from an atom smashing experiment, had found its way into a molybdenum atom and transformed it into the missing element 43?

Trying not to reveal too much, Segrè asked if he could have a look at some of the molydenum scraps Lawrence was throwing out. Weeks later a few strips of worn-out molybdenum arrived at Segrè's lab. Analysis revealed the presence of tiny amounts of elusive element 43. Segrè and Perrier announced their discovery in 1937 and named their manufactured element technetium, from the Greek *technikos* meaning artificial.

Now there was a means of synthesising technetium, scientists could investigate why it had not been found naturally. The answer was technetium itself. The 43 protons in every technetium atom do not sit there

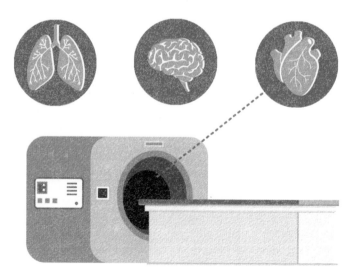

comfortably. In an effort to rearrange themselves into a more stable configuration, protons, neutrons and energy are forcefully discarded. Technetium has to be constantly remade because it falls apart so easily, releasing radioactive particles as it does so.

Different forms of technetium decay at different rates, from millions of years to a few hours. Any technetium atoms that were present when the Earth first formed decayed into molybdenum millions of years ago. What minuscule traces of the element can be found in rocks and soil on this planet are the result of geological accidents and naturally occurring nuclear reactors.

Humans have harnessed nuclear power for their own purposes, and technetium is now produced by the tonne as a by-product in spent nuclear fuel rods. Science might seem to have unleashed a monster to wreak havoc on the world. But, like the creature in Mary Shelley's science-fiction story, technetium should not be taken at face value.

Technetium is not all bad. One form, technetium 99m, decays in a matter of hours releasing gamma rays as it goes. This form of radiation deposits the least amount of energy and, together with such a short window of existence, makes technetium 99m safe to use inside the human body. The gamma rays that are emitted quickly escape but can be used to indicate the path back to the atom they originated from. Small doses can therefore be used to construct three-dimensional internal maps of the heart, brain or other organs without the need for invasive surgery. It has been used safely in millions of medical diagnostic tests.

Silver
The Flatterer

We humans can be vain. We like to be admired. We add decoration to
our bodies and our surroundings to give a certain impression. We study
ourselves in the mirror to check we are looking our best, and we take
photos so that others can remember us. Silver, perhaps more than any
other element, has pandered to our vanity. It reflects our image and
embellishes our lives. Such flattery has placed silver in high regard.

While an intellectual itch motivates many people to make new
discoveries, it is doubtful so many would have gone to so much trouble
exploring the deepest, darkest corners of South America if they hadn't
been lured there by the promise of great wealth. Tales of El Dorado
(the golden one) and Sierra de la Plata (silver mountain) tempted Europeans deeper and deeper into the continent. Though plenty of gold was
found, the hidden city where it was said to be abundant, was a myth.
Sierra de la Plata, on the other hand, turned out to be real. Cerro Rico
(rich mountain), in southern Bolivia, was accidentally discovered in
1545. It was so full of silver it provided 80 per cent of the world's supply for over two centuries and is still being mined today.

While several elements take their names from the places of their
discovery, silver has given its name to places richly endowed with the

Ag

BOLIVIA

Rio De Plata

Melting Point
962°C (1763°F)

Boiling Point
2162°C (3924°F)

Group
II

Period
5

Silver
Transition Metal

metal. Argentina, the origin of the silver mountain story, means 'made of silver'. Río de la Plata, the river that forms part of Argentina's border with Uruguay, translates as 'river of silver', for the silver that poured out of it and flooded into Europe, where it was transformed into everything from currency to cutlery and cures for warts.

Coins minted from silver physically embodied their value and resisted the wear and tear of daily use. The metal was fashioned into decorative objects to show off the owner's wealth and good taste. Polished to a mirror finish, people could use it to gaze at themselves endlessly. Silver reflects back all of the visible light that falls on it, giving it an almost colourless quality. This reflectance dips only slightly in ultraviolet light, resulting in a warm, yellow tinge that shows us to our best advantage.

Silver keeps its shine against a range of chemical attacks, but is not so nonreactive that it can't be made into useful compounds that have found yet more ways to flatter our vanity. Compounds formed with silver and members of the halogen family of elements interact with light in a special way. Rather than offering a transient reflection in a mirror, these silver compounds capture images permanently. Silver compounds were reverted to pure silver by light reflected from the faces of our ancestors. Their bright cheeks and dark laughter lines were recorded by reacted and unreacted silver on copper plates. These daguerreotypes were superseded by silver compounds coated on glass, giving us negatives that could be used to mass-produce photographs of the rich and famous. The silver screen got its name from the silver compounds layered onto celluloid that have recorded images idolized by millions.

The fine features presented in these images could be preserved using other silver compounds. From medieval times up until the twentieth century, silver nitrate was used as 'lunar caustic' to remove warts. But silver's supporting role in human health is more than cosmetic. The element has relatively low toxicity as far as humans are concerned, but bacteria and viruses are poisoned by its presence.

The Persian king, Cyrus the Great, was said to carry his own water supply with him wherever he went. Drawn from specially chosen springs, the water was boiled and kept in silver vessels. The ancient silver coins found at the bottom of so many wells may have been a practical rather than wishful gesture. Today, silver particles are embedded into the plastic of medical devices such as intravenous (IV) lines and catheter tubes to prevent infection.

Silver can seemingly do no wrong, but humans have sometimes been guilty of putting too much faith in flatterers. Attracted by its antibacterial properties, people have swallowed silver, sometimes to excess. Argyria, or silver poisoning, is caused by silver particles being deposited in the skin and eyes, turning them a bluish-grey. Its effects are distinctive and permanent, but not fatal.

Silver's allure can also corrupt. Whether it is the promise of fame from films and photoshoots, wonder drugs that cure all ills, or thirty pieces of silver, everyone has their price.

Tin
The Team Player

Tin is a fantastic team player. Combined with other metals it creates something greater than the sum of its parts. It brings out the best in others and has been an integral part of human progress. But, on its own, tin is vulnerable. The more tin is refined and separated from its fellow elements, the more likely it will break down and cry.

Over 5,000 years ago humans discovered that adding tin to copper made a much harder and more durable metal than either one separately. The alloy that was produced came to define a period of more than 2,000 years of human history. The Bronze Age transformed human society. New technologies and trade routes were established. Adding extra metals and changing the ratios of copper and tin could produce a range of bronzes each tailored for their desirable characteristics. Those in command of the raw materials and know-how to make and work bronze prospered enormously.

Towards the end of the European Bronze Age there was a man called Midas. He inherited the kingdom of Phrygia, in modern-day Turkey. In 1957 a tomb was discovered containing the skeleton of King Midas, thought to be the father of the man associated with the legend of the golden touch. The tomb contained 170 bronze vessels and dozens of

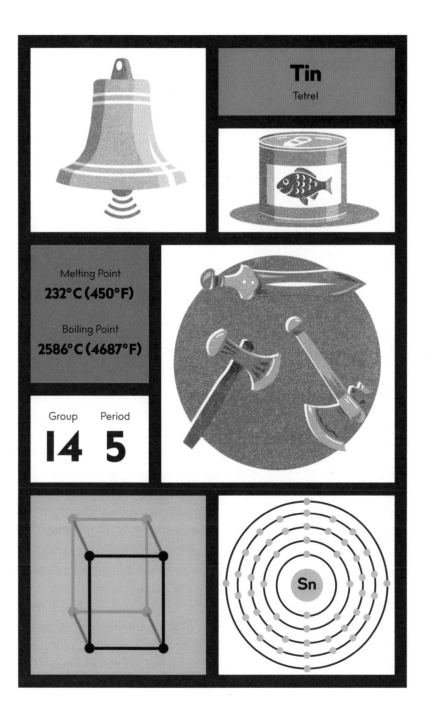

Tin
Tetrel

Melting Point
232°C (450°F)

Boiling Point
2586°C (4687°F)

Group
14

Period
5

Sn

149

decorative bronze objects. Phrygia was rich in copper. There was also a lot of zinc, which could have been used to give their bronze a yellowy-gold appearance.

The Bronze Age was succeeded by the Iron Age, but tin's contribution to major advances in human history was far from over. In the fifteenth century, Johannes Gutenberg, a German printer and publisher, had a brilliant idea. By making individual letter blocks that could be rearranged and reused, it would be possible to print almost endless combinations and reproductions of the written word. These blocks had to be soft enough to mould the letter symbols but tough enough that the words could still be read after multiple printings. Lead was too soft, but adding tin and a little antimony toughened it up. The resulting 'type metal' was easier to work than bronze but more resilient than the wood and clay blocks that had been used in the past. In 1439, Gutenberg was the first European to use moveable type, a technology that enabled the mass production of printed material and the dissemination of information on a scale never seen before.

Tin had shown its industrial and cultural worth, but it still had more to offer. It is the most tonally resonant of all metals, meaning tin bells sound the best. Tin was refined into ever purer states and moulded into bells and church organ pipes. But the metal had reached its limits. It wanted company. Pure tin exists in two forms, metallic 'white' or 'beta' tin, and fragile, brittle 'grey' or 'alpha' tin. The rearrangement occurs at 13°C (55°F) and can be heard as a crackle, more emotively known as the tin 'scream' or 'cry'. Cold northern European winters led to stories of exploding organ pipes. With no understanding of the atomic structure of the metal, people blamed the Devil. The answer was not prayer but to give tin what it wanted, company. Adding atoms of another element could cure 'tin disease', lowering the transition temperature or stopping the process altogether.

Knowledge of tin's vulnerability did not hinder another technological advance. To feed a lot of people well and efficiently it helps if some of their food can be stored for a long time. Keeping food perfectly enclosed, away from the taint of the air and microbes, was the answer. Metal cans could easily be manufactured to contain the food and tin could be melted at low enough temperatures to seal it in. Kept inside peoples' homes, tin cans would not get cold enough to cause any problems. However, cans taken to the extreme cold of the Antarctic, and sealed with particularly pure tin, may have contributed to the disaster of Scott's expedition to the South Pole.

Despite setbacks, tin still wasn't finished helping humans advance. The electronics revolution of the twentieth century was made possible by tin combined with lead. Together they made the solder that connected electronic components together into the increasingly complex devices we now take for granted. Tin, and its compounds and alloys, continue to find new applications. None of us knows what the future holds, but perhaps tin does.

51
Sb

Antimony
The Poisoned Chalice

A long, long time ago, as the story goes, there was a monk named Basilius Valentinus who searched for the philosopher's stone. His approach involved using an unusual substance, stibium, something the ancients had written about as possessing healing powers. Perhaps, he hoped it could also heal metals by purging them of their impurities to reveal pure gold.

Sadly, but not surprisingly, Valentinus' experiments produced no gold. In a fit of pique the monk threw the dregs of his failed attempts into a pig sty where the pigs gobbled up his mistakes. As a result, the unfortunate pigs became very sick. However, they recovered and subsequently grew very fat. Valentinus may have failed to find the philosopher's stone but perhaps he had discovered a wonderful health tonic instead. He promptly served up samples to his fellow monks, who also got sick but did not enjoy the miraculous recovery shown by the pigs. All the monks died and so, according to the story, Valentinus' mysterious substance became known as antimony, meaning 'against monks'.

Dead monks notwithstanding, Valentinus wrote *The Triumphant Chariot of Antimony* extolling the many virtues and benefits of this incredible element. The book was translated from German into several

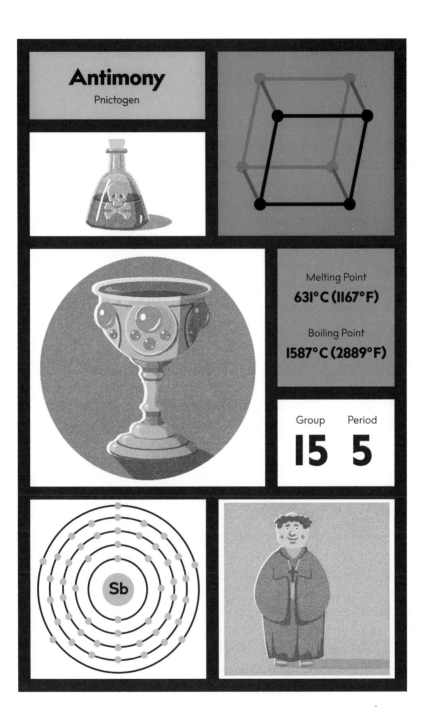

Antimony

Pnictogen

Melting Point
631°C (1167°F)

Boiling Point
1587°C (2889°F)

Group
15

Period
5

Sb

European languages inspiring yet more literature to be produced about both Valentinus and his amazing antimony. It created a medical sensation. The true author of the work was later revealed to be a German chemist and salt manufacturer, Johann Thölde. He claimed to have found the monk's Latin manuscript and merely translated it into German. Not everyone was fooled but by then it was too late, the antimony boom of the seventeenth century was well under way.

Though the story of Valentinus and his monks isn't true, it reveals a lot about how antimony and its unpleasant properties fitted right into the ethos of human health in times gone by. Up until germ theory was developed in the nineteenth century, human health was thought to be governed by the four humours: black bile, yellow bile, blood and phlegm. Too little or too much of any humour was thought to cause illness, the symptoms of which depended on whichever humour was out of kilter with the others. Balance could be restored through special diets that encouraged the body to produce more of one humour or another, or by removing excesses. The result was centuries of indigestion-inducing meals, bleeding and vomiting in the name of health. Consequently, a substance that caused someone to vomit was considered a valuable medicine. By the logic of the day, antimony was very valuable.

Antimony remedies were prescribed to the great and the good and many dramatic and high-profile recoveries were attributed to its healing powers. But there were also notable failures. Several deaths resulted from antimony-based treatments, often attributed to an overenthusiastic prescribing of these substances. Opinion on the medical benefits of the element and its compounds became sharply divided. There were the 'antimonialists' and the 'anti-antimonialists'. The authorities decided the risks were too great and banned antimony medicines, but not every physician stuck to the rule of law.

Perhaps to circumvent the rules, an alternative method of administering antimony was devised. Special pewter cups were manufactured but with a substantial amount of antimony added to the pewter alloy. Now, no medicine need be prescribed, anyone feeling unwell and in possession of a cup could fill it with white wine and leave it to stand overnight. The acid of the wine would dissolve some of the antimony from the cup and produce a potent antimony solution. Swallowing the contents of the cup the following morning would induce the vomiting that would supposedly remove the cause of the illness.

Today, with a better understanding of both the causes of disease and antimony's place in the periodic table, the element's unpleasant properties come as no real surprise. You cannot vomit away a disease and such extreme physical reactions to a substance are a clear sign of its incompatibility with good health. Antimony's position in the periodic table is also a warning sign. You can't always judge an element by the company it keeps, but you would be entirely justified in the case of antimony. Comfortably nestled in the middle of the pnictogen family of elements antimony is arsenic's bigger, nastier brother. It has similar traits to its toxic siblings but exaggerated. The riches and medical benefits antimony appeared to offer back in the sixteenth and seventeenth centuries really did turn into a poisoned chalice.

53

Iodine
The Rain Maker

Necessity, they say, is the mother of invention. Wars have often provided the necessary conditions for huge technological advances. In times of conflict, scientists and engineers are drafted in to focus their respective expertise on giving their side an advantage, be it bigger bangs or stronger defences. The collective will and collaboration between so many experts can foster an extraordinarily fertile scientific environment. And, while their eyes are always on the prize, there can be some fantastic offshoots from all this frenzied activity.

When Europe was embroiled in bitter battles, thanks to one man's ambition to dominate the continent, appeals went out to scientists to improve their side's firepower. One chemist who answered his nation's call was Frenchman Bernard Courtois. Working with limited budget and resources, he made a discovery that would benefit not just soldiers, but many thousands of ordinary citizens.

It was 1811, and the Napoleonic Wars raged across Europe. The British Navy had blockaded France to cut off its supplies of saltpetre (potassium nitrate), which was used to make gunpowder. A cottage industry in saltpetre manufacturing sprang up in France. Rotting manure and offal (to provide the nitrate) was mixed with soil and ashes

Iodine

Halogen

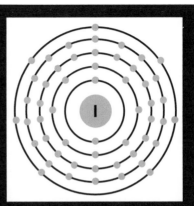

Melting Point
114°C (237°F)

Boiling Point
184°C (364°F)

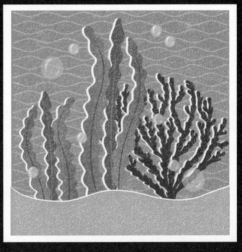

Group
17

Period
5

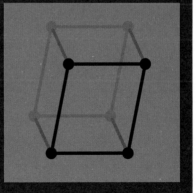

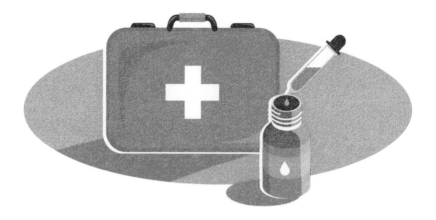

(to provide the potassium). Courtois was a saltpetre manufacturer based on the outskirts of Paris. He decided to look for alternative sources of potassium and chose seaweed because it was inexpensive and abundant. He boiled the seaweed in water to extract potassium chloride. He had achieved his main objective, but his experiments yielded a bonus discovery.

One day, he added sulphuric acid to the residues of his seaweed extractions and was surprised by strange purple fumes he saw rising from the pan. Curious, as any chemist would be, he repeated the experiment. This time he captured the purple smoke and condensed it into dark crystals that reflected a metallic lustre. Purple was an extremely difficult colour to produce at the time and had definitely never been seen arising from something that was even vaguely like a metal. It was an amazing discovery that, as a sideline to the main conflict, provoked a small scientific scuffle.

Certain he had discovered a new element, but lacking the resources to confirm it, Courtois sent samples to fellow chemists for analysis. Charles Bernard Desormes and Nicolas Clément presented the results of their investigations to the Institut Imperial de France in 1813. The strange substance was also confirmed as a new element by fellow Frenchman Joseph Gay-Lussac, who named it iodine after the Greek *iodes* meaning violet.

Meanwhile, the Englishman, Humphry Davy, a chemist held in such high regard he was allowed to visit France even when at war with Britain, also carried out tests on a sample of iodine. He, too, confirmed it as a new element and sent his report back to the Royal Institution in London. The British, unaware of Gay-Lussac's work, gave Davy the credit for the element's discovery. The disagreement over who got there first rumbled on until 1913, when everyone finally acknowledged Gay-Lussac's claim and Courtois's contribution. It took almost as long to find the benefits of Courtois's discovery.

In 1908, Antonio Grossich used tincture of iodine, a weak solution of iodine in water, in a surgical procedure. It proved to have fast-acting sterilization properties. It was later tested on a massive scale to disinfect soldiers' wounds in the 1912 Italo-Turkish War.

The majority of iodine's initial military applications were on the defensive side, protecting troops from infection. But its offensive capabilities were explored in trials every bit as unusual as the element itself. Silver iodide crystals have been used to seed rain clouds. Making it rain on demand may not seem particularly aggressive or dangerous, but the idea was to drench the enemy, bogging down their tanks and troops in the resulting mud. Tests were carried out by the British between 1949 and 1952 under the codename Operation Cumulus. One popular theory is that one test flight caused the devastating Lynmouth flood on 15 August 1952, though there is scant evidence to prove it.

Today iodine continues its military operations, but in battles against malnutrition and cancer. Iodine salts are added to food to correct dietary deficiencies that can cause birth defects and goitre, sometimes known as Derbyshire neck in the UK, thanks to the low iodine content of the soil in that region. Radioisotopes of iodine are administered to patients with thyroid tumours, destroying cancerous cells growing there. Iodine has finally found its place in war and peace.

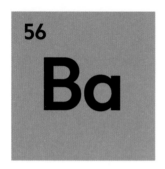

56

Ba

Barium
The Medical Marvel

Barium can cause problems, but it never means any harm. Deep down it is really just a big softy. Like Lenny in *Of Mice and Men*, barium blunders through life unaware of its own large size and oblivious to the consequences of its actions. In the story, Lenny had George by his side to keep him out of trouble, and in life barium has sulphate.

Most people will have heard of barium in the form of a barium meal. Cupfuls of this element are swallowed every day in hospitals around the world, but it is far from nutritious. In many respects, this element takes after calcium, its lighter chemical sibling, but it is big and clumsy by comparison.

One of calcium's many biological roles is to regulate the movement of potassium, generator of electrical signals, in and out of cells. Barium, stumbling down the same chemical path, does nothing more sinister than get in the way. It blocks potassium's path out of the cell. Trapped inside, potassium cannot generate the signals that keep the body coordinated. Barium's bumbling can indirectly cause cardiac irregularities, anxiety, tremors, shortness of breath and paralysis. Doctors would never risk using something that could cause these symptoms in a patient, but that's where sulphate comes in.

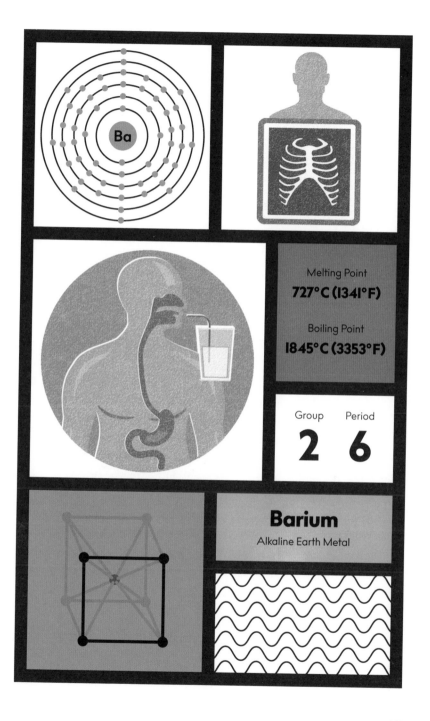

Melting Point
727°C (1341°F)

Boiling Point
1845°C (3353°F)

Group **2**   Period **6**

Barium
Alkaline Earth Metal

Barium shares the same generous spirit as calcium, giving away two electrons to any element that wants them to form partnerships. Not every friend is supportive, however, and some can lead us astray. Some partners simply can't keep hold of barium, for example, letting it wander off and get into trouble. But barium and sulphate are besties. They are so well matched they can scarcely be separated. This pair will stick together through thick and thin, even through the acid and alkali of your digestive system. With sulphate keeping it safe, barium's weighty characteristics can be turned to an advantage.

The defining difference between one element and the next is a single proton, and to make a complete, balanced atom these protons come with extra electrons and neutrons. As you move through the periodic table, these tiny incremental additions build up. By the time you get to barium, things are starting to get a little unwieldy. This metal has 56 protons and, usually, 82 neutrons packed into its nucleus. It's quite a lot of mass to be carrying around. The 56 electrons that swarm around barium's nucleus pad out its not inconsiderable girth.

A bit of extra padding can have benefits when it comes to X-rays. The more electrons an atom carries, the more easily it can absorb the X-ray's energetic blows. It is why lead, a real heavyweight with 82 electrons, is used to make screens for radiographers to stand behind to protect them from the X-rays they work with day in, day out.

Barium, despite its name deriving from *barys*, the Greek word for heavy, ranks as a middleweight by comparison. But, it has enough bulk to absorb X-rays and reveal the squishy contours of your digestive system. It is the preferred ingredient for the sludgy diagnostic dinners. Swallowing a slurry of lead compounds would show up your insides more effectively, but its toxicity would kill you. Our bodies, chemically speaking at least, are almost completely unaware of barium sulphate as it passes through. Biology, for the most part, bypasses barium, but there always exceptions.

There is one group of organisms that can't get on without it. Members of this small but spectacular group of algae are called desmids. These little, green, living snowflakes grow up to a millimetre in size and are characterized by their divided structure. They all have two lobes connected at a central point. The lobes can be spherical, spiked, smooth or knobbled, but they all have fluid-filled pockets built into their extremities. Along with the liquid sloshing around in these pouches are insoluble crystals of barium sulphate. Though these barium sulphate crystals have no chemical interactions with their surroundings, they are constantly moving around, buffeted by the water molecules that surround them. The exact purpose of the crystals is unclear but one theory is that the heavy barium in the crystals acts like a little gravity sensor helping the desmid figure out which way is up. Barium might not be the most dynamic element, but with the support of sulphate, it is a safe and reliable friend.

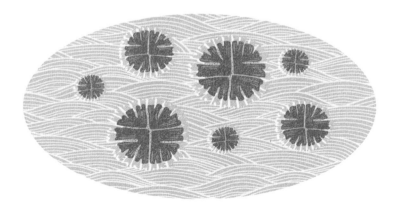

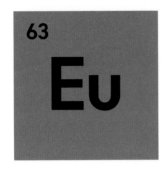

63

Eu

Europium
The Forger's Foe

Europium is exasperating. It is one of a large group of near-identical elements that confounded chemists for more than a century. Now this ability to perplex has been turned into an advantage. Europium's confusing qualities can be used to spot fakes and deter forgers.

Many groups in the periodic table have strong family traits, but some elements are so similar that distinguishing between them is like playing a particularly tricky game of spot-the-difference. One group of seventeen elements behave in such a chemically similar way it makes sense to lump them together, even though they are located in different parts of the periodic table. Known as the rare earth elements they are neither rare nor earths. A more refined categorization of these chemical doppelgängers has been made based on how the electrons are arranged within their atoms. These are the lanthanides, a fifteen-member subset of the rare earths spread in a long line beneath the main periodic table. The group includes lutetium, even though it shouldn't because lutetium's electron arrangement means it is technically a transition metal and should be part of the main table.

Confused? You aren't the only one. The chemical similarities between the rare earths means they are usually found together in the same rocks.

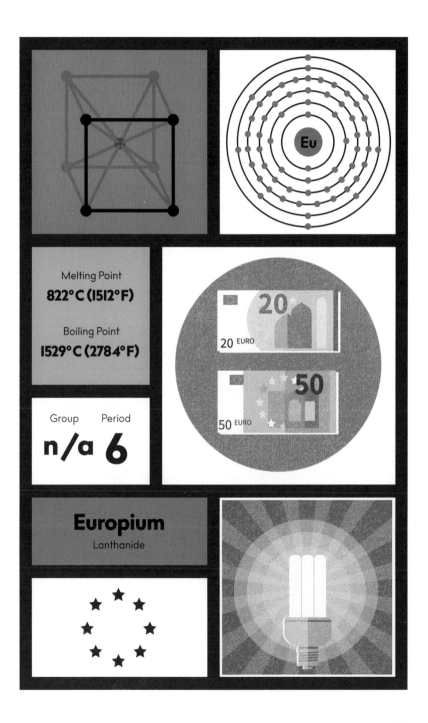

Melting Point
822°C (1512°F)

Boiling Point
1529°C (2784°F)

Group
n/a

Period
6

Europium
Lanthanide

Over the course of the nineteenth century chemists tried to unpick the tangle of elements. Seemingly every time one element was isolated, another one was found hiding within it. Arguments broke out over who had discovered what and when. It was a nightmare.

Even William Crookes was at his wit's end and, as the discoverer of two elements (helium and thallium), he knew what he was complaining about. 'The rare earth elements perplex us in our researches, baffle us in our speculations, and haunt us in our very dreams. They stretch like an unknown sea before us, mocking, mystifying and murmuring strange revelations and possibilities.'

Towards the end of the nineteenth century, europium was found hiding as an impurity within samarium. French chemist Eugène-Anatole Demarçay decided to try and isolate it. Taking advantage of the very subtle differences between samarium and the as yet unknown element, he invented a method of separating them by a series of painstaking reactions and recrystallizations. It took him years. After so much hard work no one had the heart to contest his claims to discovery.

What is all the more soul-destroying is that it needn't have taken him so long. Europium is unusual amongst the rare earths in that it can readily give up two or three of its electrons. The remaining members of the family are less flexible and usually only offer up three electrons. Chemically, it makes all the difference. Europium can make a range of compounds inaccessible to most other lanthanides, making it easy to separate from its siblings.

In most other respects, however, europium is just like all the other rare earths. Few people bother to distinguish between the different elements in the family. Some may be more valuable than others because they are less abundant, but they are usually sold together as a mixture called 'mischmetall'. Nevertheless, there are a few specialist applications that make it worth the trouble of separating individual elements from the rare earth mix. One of those applications is spotting forgeries.

All lanthanides, because of how their electrons are arranged, fluoresce. Shining certain wavelengths of light on their atoms causes their electrons to jump to higher levels. When those electrons fall back to their original arrangement they re-emit the light they absorbed, but at a different wavelength. A blue light can therefore shine back as red. Different colours can be produced depending on which lanthanide is used and what it is mixed with. It is one of the many tricks used to make bank notes difficult to forge. Holding real bank notes under an ultraviolet light will reveal hidden colours and patterns printed with lanthanide compounds.

In 1995 a new currency was created. When it came to designing the notes, any lanthanide could have been chosen to make glowing security features. They probably chose europium because its florescent properties are well understood, and it has a long history of use in TV displays and fluorescent lamps. Perhaps cost was another factor, because being more expensive than many other lanthanides, it would deter forgers who wanted to maximize their profits from their fakes. But, when the currency is called the euro, only one lanthanide will do.

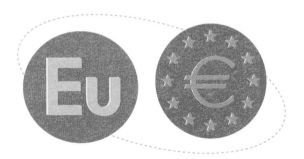

Tungsten
Bringer of Light

Tungsten is the tough-as-nails heavyweight of the periodic table. It radiates an aura of solidity and imperviousness. You would always want tungsten on your side in a fight, but tungsten, by its very nature, is not easily persuaded to do anything it doesn't want to do. Tungsten goes by different names and has been used in defensive and offensive roles, but it is always prized for its indifference to what the world throws at it.

Many of tungsten's qualities were apparent from the start. A stone found in Sweden in 1751 was described as *tung sten*, literally 'heavy stone'. Intrigued, the great Swedish chemist Carl Wilhelm Scheele decided to take a closer look. He found the white crystalline material within the stone was a combination of calcium and something else. He could produce an acid, called tungstic acid, from the something else and he suspected this acid contained a new metal. That metal would presumably be named tungsten as and when someone managed to isolate it.

Two years later, and several hundred miles away in Spain, the brothers Fausto and Juan José d'Elhuyar did isolate Scheele's mystery metal, but from a different mineral, wolframite, so they named it wolfram. Therefore, the element has the letter symbol W, though the element

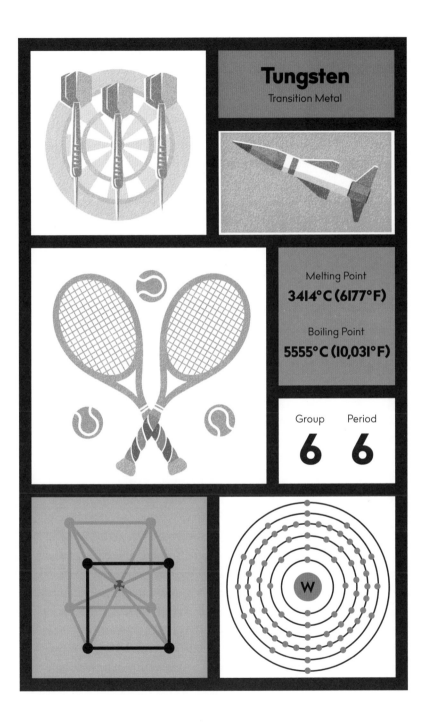

Tungsten

Transition Metal

Melting Point
3414°C (6177°F)

Boiling Point
5555°C (10,031°F)

Group
6

Period
6

W

is called tungsten in several languages. However, the Spanish, understandably, call the element *wolframio*. The Swedes, despite their man Scheele's contribution to its discovery, call it *volfram*.

Whatever you call it, it's heavy. Very heavy. So heavy, lead feels like a lightweight by comparison. And unlike soft, malleable lead, tungsten is tougher than steel and keeps its strength even at high temperatures. It eventually melts at a staggering 3400°C (6150°F), the highest of any metal. The shear heft and inflexibility of this element made it an attractive material for military applications. Turned into ammunition, the huge amount of kinetic energy a tungsten projectile could carry with it would cause enormous damage to anything it was fired at. In the early nineteenth century, excited at the prospect of producing a formidable weapon, people made efforts to manufacture tungsten cannon balls, but the metal was simply too difficult to work.

For a long time tungsten remained stubbornly indifferent to the chemical, physical and energetic powers of persuasion that were brought to bear on it. As new techniques and treatments were developed, tungsten slowly submitted to the scientist's will. Surprising applications were found for tungsten and its compounds, and all played to its strengths.

In the Victorian era combustion was a considerable problem. Homes were illuminated with the flames from oil lamps and candles, and women's dresses were a mass of highly flammable material. Tungsten offered solutions to both problems. Cloth treated with sodium tungstate made the material incombustible and tungsten metal was the ideal choice for the incandescent lights bulbs that were to eventually replace candles and lamps. Forcing an electric current through a fine metal wire would heat it up to the point where it would glow. Tungsten's resilience at high temperatures made it the clear favourite.

But, even tungsten is not impervious to the extreme temperatures needed to make it glow. Early bulbs blackened as the metal evaporated and deposited on the inner surface of the glass. Subsequently, lighter, denser coils and an inert gas pumped into the glass bulb protected the tungsten from evaporation and corrosion. With these modifications, tungsten bulbs came to dominate domestic lighting.

Then, just as more and more homes were filled with the cheerful glow of hot tungsten, the lights suddenly went out. The Second World War enforced blackouts across Britain. Windows were covered and lights put out to hide from the German bombers flying overhead. But, tungsten was more in demand than ever. Military technology may have moved on from cannonballs but tungsten's heavyweight credentials made it a highly coveted element for ammunition. Tungsten-tipped missiles – named kinetic penetrators – could stop a tank. Adding tungsten to steel also made excellent machine tools that aided the manufacture of military vehicles and weapons.

Today, tungsten has found a role in different battles – in the field of sports rather than war. It remains desirable for its strength and weight. The metal is used to make high-quality darts and tennis rackets. Manufacturing still requires good-quality machine tools whether it is tanks or turbines that are being made. The manipulation of tungsten has become more sophisticated, but it is still the same brute qualities being used, just in ever more refined ways.

Platinum
Little Silver

For a long time, platinum was the underdog. Its name, little silver, reflected its lowly status. Gold prospectors who found nuggets of pure platinum in their gold pans would throw it back into the deepest part of the river so it would not trouble them again. South American gold mines were abandoned when platinum was found contaminating the deposits. How things have changed.

Platinum simply got off on the wrong foot, at least as far as Europeans were concerned. The Spanish had discovered large deposits of gold in South America, but they were often contaminated with another metal. This metal looked like silver – hence *plata* the Spanish word for silver – but it was heavy, difficult to work with and got in the way of much more valuable gold – hence *ina*, meaning little.

For the gold-obsessed Spanish, *platina* was a huge inconvenience. It was more or less thrown away as waste. But one person's inconvenience is another person's opportunity. Platinum had a very similar density to gold and was therefore used to adulterate gold coins. The Spanish authorities were outraged that their currency was being devalued with cheap fakes and they ordered any counterfeit coins to be sunk to the bottom of the sea. It was not a good start and things did not get much better for a long time.

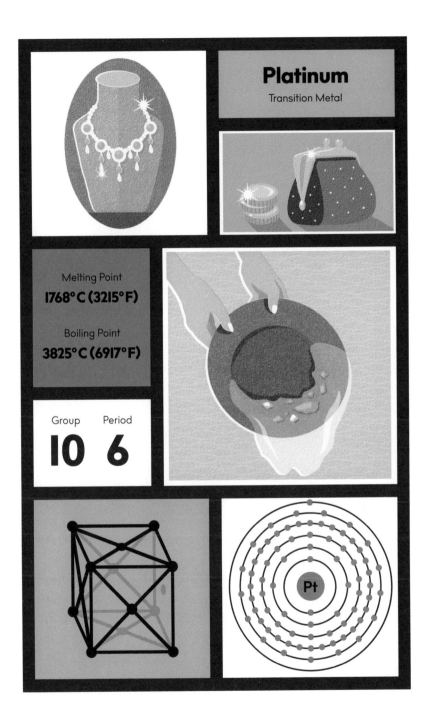

Platinum
Transition Metal

Melting Point
1768°C (3215°F)

Boiling Point
3825°C (6917°F)

Group
10

Period
6

Pt

Around 1750 platina was introduced to Europe as a curiosity. No one was sure what to do with these tough nuggets and so the French chemist Pierre-Françoise Chabaneau was set to work to make a malleable form of the metal. He gradually freed platinum from the naturally present impurities that made platina so intractable. It was a frustrating process and, at one point, Chabaneau smashed all his laboratory equipment shouting, 'You shall never again get me to touch that damn metal!' Three months later he revealed a 10cm (4in) cube of pure, malleable platinum. Ingots of the pure metal were shown to the aristocracy of Europe, who were intrigued by its similarity to silver and stunned by its incredible weight.

The fact that the new metal was more resistant to tarnishing than silver, and cheaper, made it a clear substitute for decorative pieces. But platinum stubbornly resisted attempts to bend it to the metalworkers' will. Temperatures of over 2000°C (3600°F) are needed to melt platinum so it can be poured into moulds – well beyond the range of charcoal fires. It is also much stronger than silver, so it cannot easily be beaten into shape or decorated with patterns etched into the surface.

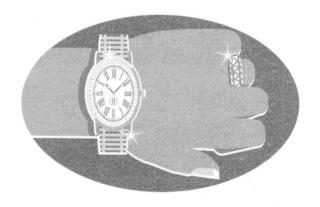

None of this would have been a surprise to the indigenous population of South America who had known about platinum for at least 2,000 years before the Europeans arrived. They were well aware of platinum's decorative potential and difficult behaviour, but they had found ways to manipulate the uncooperative metal and had been fashioning it into jewellery for millennia. It is thought they used a process called sintering. Small pieces of platinum and gold could be mixed then heated together and they would slowly meld and merge into one solid piece of metal. The platinum-gold alloy would then be repeatedly beaten and heated to remove the gold. Eventually, and independently, European scientists found ways to manipulate platinum, but it was such hard work there seemed to be little to gain. However, it was this strength and hardness that was eventually to come to platinum's rescue.

Towards the end of the nineteenth century Louis Cartier took the decision to use platinum whenever he could in place of silver and gold in making his much sought-after jewellery. Gold settings could look gaudy next to precious stones. Silver showed off the colour of gems better but eventually tarnished and both metals were relatively soft, meaning bulky settings were needed to hold the stones securely. Platinum was a perfect alternative. Tiny but robust mounts could be made out of this strong metal. Its white colour and soft shine never dulled and showed off the brilliance of diamonds wonderfully.

Platinum's stock rose throughout the twentieth century primarily because of its use in jewellery. Slowly it came to be associated with class and good taste. In the 1930s it became part of everyday fashion when Jean Harlow dyed her hair platinum blonde. When gold cards and gold records become blasé, people searched around for an element that spoke of even greater wealth and prestige, and they hit upon platinum. Finally, the underdog had triumphed.

Gold
The Nanotechnologist

Gold is so iconic it has become more than an element. It is not just a metal or another square in the periodic table, it is a colour, an adjective, a symbol of wealth, and a mark of quality. The element has been known and prized since antiquity, but what makes it so special? Its warm colour, soft lustre and malleability make it highly desirable. It can be moulded and bent into attractive shapes and patterns that never tarnish. Even disguised or in tiny amounts it can add an extra wow factor to an object. Gold is. . .well, it is the gold standard.

For thousands of years gold was thought to be not just the best, but perfect. All other metals were inferior by comparison because they became tainted, a sign, it was thought, of their impurity. Some thought these other metals would be transformed into gold if only these impurities could be removed. Many tried but none succeeded. Gold is gold because it has 79 protons. The only way to transform one metal into another is to tinker with the nucleus, adding or removing protons, something that only became possible in the twentieth century using particle accelerators.

In the past, some reasoned that if gold was free from taint, it might rid the human body of impurities, too. It might offer medicinal

Gold

Transition Metal

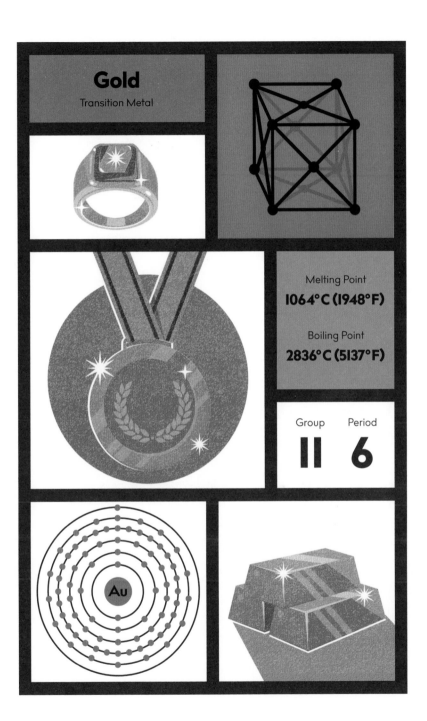

Melting Point
1064°C (1948°F)

Boiling Point
2836°C (5137°F)

Group
11

Period
6

Au

benefits, if only there was a way of introducing it into the body. Gold flakes or dust could be sprinkled on food, but better still would be to create a liquid form of gold that could be drunk. Gold melts at temperatures over 1000°C (1800°F) and swallowing molten gold would kill rather than cure. So, imagine the delight when a liquid was discovered that could dissolve gold.

To a medieval alchemist, a liquid that dissolved gold – a metal famed for its imperviousness – was something very special, so they named it *aqua regia* or king of waters. The recipe had been taken from a much older text written by the seventh-century Muslim scholar Jabir ibn Hayyan. It called for all manner of elaborate ingredients, but in fact boiled down to a potent mix of hydrochloric and nitric acid. Clearly, swallowing this, even with gold dissolved in it, was a recipe for disaster. However, alchemists found that the gold solution could be diluted with rosemary oil without releasing the gold, but without any therapeutic benefits either. Gold therapy made a comeback in the twentieth century when scientists found two gold compounds, sodium aurothiomalate and auranofin, that could be injected to treat rheumatoid arthritis.

There is another example of an ancient use of gold that is also very modern. The Lycurgus Cup dates from the fourth century and it is extraordinary. Made from glass, it shows King Lycurgus entangled in vines, all brought into relief because the surrounding glass has been carefully ground away. This alone would make it a magnificent

Gold

artefact but there is more. The cup can change colour, and all thanks to tiny particles of silver and gold that take up just one per cent of the volume of glass. When light shines on the cup it appears green, like jade, caused by reflections from silver particles. But when light is shone through the cup it turns red, because of gold. The ratio of silver and gold had to be spot on. The temperature of the furnace had to be perfect. A few other impurities were also needed to reduce the metals to the ideal state.

The technology was forgotten again until the seventeenth century, when gold dissolved in aqua regia was added to glass to give it a ruby or cranberry colour. But thirteen hundred years later they could not reproduce the colour-changing trick. The explanation for that came in the twentieth century. It turns out the Roman glassmakers had been dabbling, albeit unintentionally, in the very modern science of nanotechnology – the use of nano-scale substances. What probably happened was that electrum, a naturally occurring alloy of silver and gold, was added to a batch of ordinary glass to make it coloured. Then a lot of coincidences contrived to produce gold and silver particles between fifty and one hundred nanometres, or billionths of a metre, across, which is just the right size to interact with visible light and cause the 'magical' colour-shifting effect.

Many believe nanotechnologies promise a better, greener, more efficient future. Gold, all those years ago, was setting another standard.

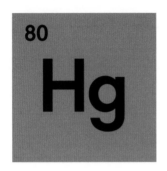

80

Hg

Mercury
The Weirdo

Mercury is the stuff of legend. It has made hatters mad, killed kings
and obsessed alchemists. Speculative scribblings about this strange
element stretch back thousands of years and fill hundreds of volumes.
Scientists, philosophers and the curious have devoted hours to contem-
plating this enigmatic metal. What is it about mercury that generates
such interest? Well, mercury is deeply, and unapologetically, weird.

There are many things in science that, although undoubtedly correct,
seem to defy reason. We all try to make sense of the world by compar-
ing unknown things with known things in an effort to predict how
they might behave or what to expect of them. But mercury is not really
like anything else. It is shiny like silver, but liquid like water; lead floats
on its mirror-like surface, but gold disappears into its depths. Mercury
makes no sense.

Mercury's strange behaviour gave it a unique status in the eyes of
alchemists. The way the metal moved and skittered across surfaces
made it seem almost alive. Mixed with gold and heated at the right
temperature over several days, mercury can sprout growths that look
for all the world like a living tree. Its unusual properties were suspected
of being hints of some deeper reality or fundamental substance, and for

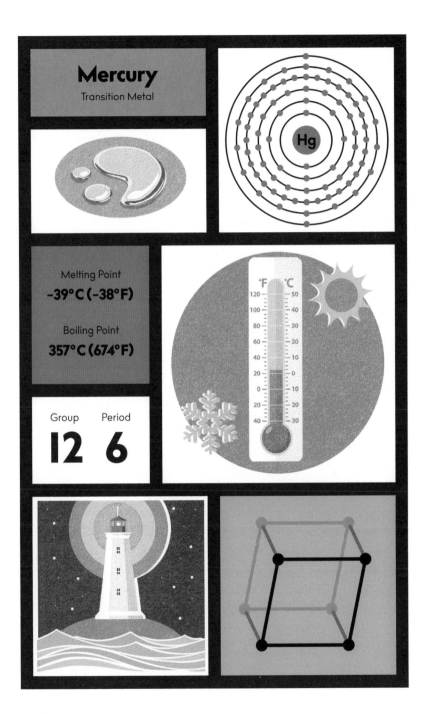

Mercury
Transition Metal

Hg

Melting Point
–39°C (–38°F)

Boiling Point
357°C (674°F)

Group
12

Period
6

centuries alchemists threw everything they could at this strange material to try and tease out its hidden secrets.

Many thought mercury might be, or contain, the very essence of matter. If it could be extracted or used in some way it could, perhaps, confer that essence to other things. Mercury was often central to the search for the philosopher's stone, the mythical substance that would magically transform any metal into pure gold. It had equal importance in the hunt for an elixir of life, a medicine that would cure all diseases and extend life indefinitely. Such speculations were not confined to the gullible or greedy. Great scientists such as Isaac Newton experimented with mercury. King Charles II of England was poisoned by the mercury experiments he carried out in a poorly ventilated room in his palace.

But there is no philosopher's stone, and the secret of eternal life still eludes science. As the eighteenth century and the Age of Enlightenment progressed, alchemy looked increasingly outdated and unreliable. Mercury, however, never stopped being strange. Even in an increasingly scientific world, many refused to accept it was even a metal.

When travellers returned to Europe from Siberia with tales of mercury freezing into a solid like any other metal, they were dismissed

as fanciful stories. Mercury was too extraordinary to be so convention-al. It was the embodiment of fluidity, hence its other names quicksilver and *hydrargyrum* ('silver water' in Latin and from where it obtains the letters of its chemical symbol Hg). But, in December 1759, two Russian scientists, Joseph Adam Braun and Mikhail Vasilyevich Lomonosov, changed all that.

The pair were conducting experiments into extreme cold. Mixing snow with salt lowered the temperature to below zero, as shown on their mercury thermometer. By adding acids and other substances they could force the temperature lower still. Eventually the mercury in their thermometer stopped falling. Perplexed, they broke the glass to reveal a solid sphere of mercury, where the thermometer's bulb had been, with a thin wire protruding from the top that bent just like ordinary metal.

Stripped of their mystical sheen, the characteristics that made mer-cury so strange could be exploited like any other material. If mercury had such a liking for gold, it could be used to extract the precious metal from ores. If a substance so damn heavy was going to insist on being a liquid, then it could be used to float the weighty lamps inside lighthouses so they would spin at the gentlest push. If it really was a metal, it could be used in electronics, stopping and starting the flow of electricity as the liquid sloshed about inside switches.

These applications have been relatively short-lived. Mercury is, after all, mercurial. Such usefulness comes at a price, and that price is health. Though the dangers have been known since ancient times, such high costs are no longer tolerable. Mercury's toxicity means it has been replaced wherever possible. Nevertheless, the only metal in the periodic table to be liquid at room temperature still fascinates us today because it is unlike anything else. It does not conform or compromise, and flat out refuses to be pigeon-holed. Mercury, weird though it may be, is proudly and unashamedly nothing but itself.

Thallium
The Wolf in Sheep's Clothing

To say that Fidel Castro was disliked by the US government would be an understatement. At least eight administrations are alleged to have plotted the demise of the Cuban leader. Over forty years, the CIA made 634 assassination attempts, according to the former head of Cuban intelligence, Fábian Escalante. In their efforts to eliminate El Comandante it would seem every conceivable means of maiming or murdering were considered, from the sublime to the ridiculous.

One idea explored in the early days of Castro's regime was to spray him with an LSD-like substance while he was giving a speech. The plan was to shame or humiliate the leader rather than kill him, but things escalated quickly. By August 1960 a proposal to poison Castro's favourite cigars got as far as contaminating a box of fifty with the potently poisonous botulinum toxin. Fidel would not even have to smoke them, just putting one cigar to his lips would be enough to kill him. However, the cigars ended up in the CIA deputy director's safe rather than on their way to Cuba. Other plots, including seashells packed with explosives and diving suits dusted with disease-causing fungus, were abandoned because they were either impractical or thwarted before they could get off the ground.

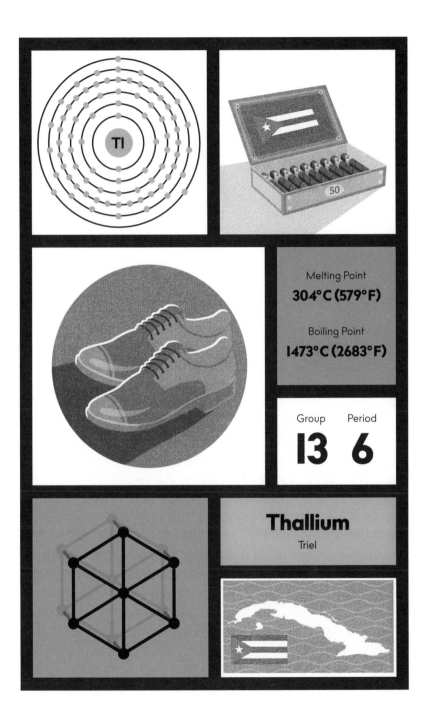

TI

Melting Point
304°C (579°F)

Boiling Point
1473°C (2683°F)

Group
13

Period
6

Thallium
Triel

However, when the CIA learned that Castro was planning a visit to China, they saw a golden opportunity to put yet another plot into action. What they needed was something discreet, something that would humiliate while it killed, but something that was difficult to detect or at least hard to trace back to the CIA. That something would have to be particularly devious to meet such a tall order, and so they looked to thallium because there are few things more devious than this element.

Thallium operates like a secret agent. The salts of this metal can enter the human body via the skin, the lungs or the digestive tract without raising the alarm. It takes on an assumed identity, a disguise to fool the casual observer, or the biological protections within the human body, at least for a while. Thallium presents itself as ordinary potassium. It works its way into potassium's positions in the body only to disrupt potassium's normal functions. Thallium can also bond to sulphur atoms, twisting sulphur-containing enzymes and proteins into shapes that do not work so well.

A plan was hatched to add thallium salts to Castro's shoes when he put them out to be shined. The poison would take effect slowly and the symptoms it caused would depend on what potassium function had been disrupted and which enzymes were corrupted. It would puzzle doctors and disguise the true cause of the Cuban leader's ill health. One common symptom of thallium poisoning that particularly interested the CIA was hair loss. Not only would the poison in his shoes make him very sick, perhaps even to the point of death, but it would make

Thallium

his famous beard fall out, humiliating him in the process. Thallium was also easy for the CIA R&D department to obtain. In the 1960s, thallium salts were sold over the counter in depilatory creams, for cosmetic purposes or to clear the way for treating skin conditions.

The scheme got as far as procuring the thallium and testing it on animals. Why such testing was needed is not obvious, the poisonous effects of thallium were well known at the time and it was the key ingredient in commercial rat poisons and insecticides. The fact that it caused hair loss was a clear sign of its detrimental effects on the human body, even if at the time it was seen by some as a positive quality. Perhaps the scientists wanted to test the method of delivery. Whether tiny little thallium-laced shoes were made for the animals to wear is not recorded, at least not in the official documents that have been released so far. Then Castro cancelled the trip.

The thallium plans were shelved, but over the following decades hundreds more ideas were tested, tried and failed. Castro died in 2016 at the ripe old age of ninety. Though no cause of death was given, and his body cremated the morning after his death, it hardly seems likely the CIA was behind it. The last images of him, meeting world leaders just a few weeks before his death, show him with a full head of hair and his trademark beard, white with age, but still intact.

82

Pb

Lead
The False Friend

Lead is the Iago of the periodic table. You think it is your friend but it is not. To your face it is always supportive, charming and reliable. But, behind your back, it is plotting your total destruction and manipulating your world to its own nefarious ends.

Unlike some other metals, lead could never rely on its looks or social standing to inveigle its way into the human world. It has none of the glamour or prestige of some of its fellow metals, it is dull in finish and colour. Lead makes you think you need it in your life by being helpful. It can feel as though lead always has a hand at your back to support you. In the past, not everyone noticed the hand was holding a knife.

For thousands of years humans were fooled by lead's usefulness and were tricked into making the most of its excellent properties. On its own lead is very soft, easily malleable and resistant to corrosion. It was perfect for forming pipes and a hundred other useful shapes. Lead enabled the Romans to elevate their society to whole new levels of prosperity and health through shifting enormous volumes of water to support the huge population within its vast empire.

Lead compounds also appeared to offer qualities humans desired. The Romans made many technological advances, but they still lacked

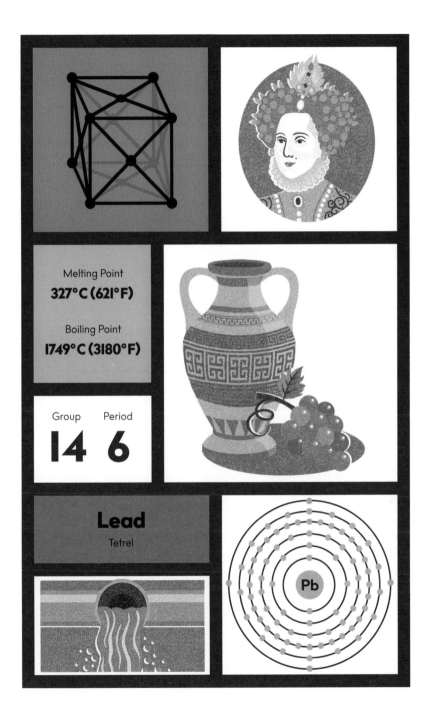

Melting Point
327°C (621°F)

Boiling Point
1749°C (3180°F)

Group | Period
14 | **6**

Lead
Tetrel

Pb

in some areas. Sugar was unknown to them. Honey was all they had to sweeten their food. That was until lead offered its services. Lead acetate, also known as sugar of lead, has the unusual quality of tasting sweet. Romans boiled the grape mash, left over from wine making, in lead pans. The acetic acid of the grapes reacted with the pan and lead acetate dissolved into the mash to make 'sapa'. Sapa was then added to sour wines to make them exceptionally palatable.

If lead had aimed to become indispensable, it certainly achieved it with the Romans. But lead was not finished with humanity yet, not by a long shot. In the sixteenth century, the ageing Queen Elizabeth I of England sought to preserve her youthful looks for as long as possible. Layers of white lead were plastered onto her face to cover her pockmarked skin and disguise wrinkles and other blemishes. She set a fashion that continued for centuries.

Even the modern world, with the benefit of greater scientific understanding, could still be taken in by lead's abilities. In the twentieth century, cars promised to transform society, but the engines that powered them stuttered and faltered when burning pure petrol. Tetraethyl lead came along to save the day. In tiny quantities it modified the way petrol ignited and saved engines from knocking themselves to a standstill.

Lead has helped humans to build great empires, given them the appearance of youth and smoothed their path to the modern world. But

at considerable cost. All along, lead found its way inside human bodies, in food and drink, through their skin and in the air they breathed. While it gave every appearance of being helpful, the element was actually doing great harm. It damaged nerves and kidneys. The signs were there from the start, but many chose to ignore them, or claimed the obvious harm to their bodies was desirable.

Roman prostitutes who ate sapa by the spoonful did so because it made their complexions pale. Lead was damaging the enzymes normally used in the production of red blood cells, making them weak and anaemic. It also made them sterile; to modern eyes this is evidence of poisoning, but was seen as an added bonus at the time. The lead in everyday water, food and wine slowly poisoned the Roman population, to a greater or lesser extent. Lead may not have brought about the collapse of the Roman empire single-handedly, but it certainly helped. Despite knowing the potential dangers of lead, Renaissance women were not deterred from its daily use on their skin. It blackened their teeth and made their hair fall out. Fashions merely adapted to the damage. High foreheads and black teeth were considered beautiful.

No matter how great the benefits or small the quantity, lead is not going to help us in the long run. Lead compounds have been stripped from cosmetics, paint and petrol. The substitutes may not perform quite so well, but they are much less likely to do us harm. It has taken us a long time, but humans have finally realized that lead simply cannot be trusted.

Bismuth
The Looker

Most elements can be judged by the company they keep, but not bismuth. It is surrounded by some of the nastiest, most toxic elements in the periodic table, and you would expect it to be just as unpleasant. However, at the bottom of the group known as the pnictogens, sometimes referred to as 'poisoners alley', lies a spark of hope and happiness.

Bismuth is the Dorian Gray of the periodic table, beautiful and untouched by the passage of time. Despite being associated with poisoners, and having nasty habits that include radioactivity, bismuth maintains its good looks and excellent reputation. And there is no ugly portrait in bismuth's attic.

Few metals are admired for their inherent beauty in the way that bismuth is. Many people admire sculptures moulded and shaped out of different metals, but few metals are displayed as ornaments in their own right. However, crystals of pure bismuth adorn many desks and bookshelves because they grow in stunning step-like patterns that reflect rainbow colours like an Escher drawing on an acid trip. These colourful inside-out pyramids are created because the crystal forms fastest at the edges, leaving angular hollows. The iridescent finish is from white light diffracted into different colours through layers of bismuth

Bismuth

Pnictogen

Melting Point
271°C (521°F)

Boiling Point
1564°C (2847°F)

Group
15

Period
6

oxide that coat the surface. When most metals react with oxygen, they are usually left dull and tarnished but bismuth becomes even more beautiful. Bismuth looks great, and it can make you look great, too.

Combined with oxygen and chlorine, bismuth forms bismuth oxychloride, a compound also known as pearl white because of its resemblance to mother-of-pearl. The layers formed by the oxygen, chloride and bismuth reflect light at different depths throughout the crystal and give it a beautiful pearl-like shine. The ancient Egyptians added it to their cosmetics for a bit of extra glamour. They started a make-up trend that continues to this day and has seen bismuth oxychloride included in hair and nail products, as well as eyeshadows and face powders. The thin flakes of pearl white reflect and scatter light giving skin a soft glow and blurring out creases and defects.

But is it wise to put heavy metal compounds on your face? Especially when that heavy metal comes from a long line of notorious poisoners. Fear not. Bismuth is the black sheep of its disreputable family because it is not just safe but, in some circumstances, it is even good for you. Pepto-bismol, the popular antacid, kills off the bacteria that cause stomach upsets and soothes inflamed stomach linings (but always read the label first). If bismuth sounds too good to be true, surely a bit of radiation could dent its reputation.

From its position in the periodic table, bismuth should be unstable and disruptive, but instead it is remarkably content and self-restrained. The atoms of every element from hydrogen onwards must pack more and more positive protons into their tiny nuclei. Positive charges repel each other, so to stop everything flying apart, increasing numbers of neutrons also have to be included to hold everything together. This system works, more or less, all the way up to lead. Elements heavier than lead have so many protons and neutrons that they become unmanageable. To lighten the load, these heavier elements eject particles from the nucleus, at various rates and with varying force, to transform themselves into lighter, more stable elements. This is radioactivity.

Bismuth is one step on from lead, the last stable element in the periodic table, and it should therefore be radioactive. Theoretical calculations showed it should indeed decay by spitting out alpha particles (units of two protons and two neutrons) to become stable thallium. And though bismuth is radioactive, it goes about it in such an understated way that it was missed for decades.

It was only in 2003 that scientists spotted the elusive metamorphosis and detected the alpha particles that were ejected by bloated bismuth nuclei. In order to do so, the scientists had to gather a lot of bismuth atoms (sixty-two grams worth, or roughly 178 million quadrillion atoms), place them inside an extremely sensitive detector and wait. Over five days just 128 bismuth atoms spat out alpha particles to become thallium. For half the atoms of bismuth in their sample to decay would take a billion times longer than the age of the universe. Bismuth is so discrete in its radioactivity it might as well not be radioactive at all. It is the most stable unstable element in the periodic table. Bismuth will outlast us all.

Polonium
The Devil Incarnate

There is little to love about polonium. It is one of the deadliest elements in the periodic table and has few redeeming features. The small number of uses that have been found for the element are uncommon and very specialized. It would have vanished into well-deserved obscurity but for its association with international political disagreements.

The element was discovered by Pierre and Marie Curie in 1898. They were intrigued by samples of uranium ore that were more radioactive than should be possible from the uranium alone. They were motivated to hunt out the source of the intense radiation, suspecting it was a previously unknown element. It took years of sifting and refining tonnes of uranium ore residues. Their work left them with hands scarred from radiation burns, ruined health and a few hundredths of a gram of, not one, but two new metals. They had already decided on names for these new elements; one would be radium, because of its glow, and the other polonium, after Poland, Marie's homeland. At that time the country had been annexed by neighbouring states. The Polish language and culture were heavily repressed.

Marie's plan was well intentioned, but the association did not benefit from closer scrutiny. With hindsight, it was not the most flattering

Polonium
Chalcogen

Po

Melting Point
254°C (489°F)

Boiling Point
962°C (1764°F)

Group
16

Period
6

POLAND

honour to bestow on a country struggling to achieve its independence. Polonium metal was next to useless and terrifyingly hazardous to human health. The Curies, however, were understandably thrilled with 'their elements'. Some evenings, the couple would return to the lab to admire the glow emanating from their hard-won samples.

In 1904, when excitement about the new elements was still high and relatively little was known about their dangers, Mark Twain included the Curies' discoveries in a short story, 'Sold to Satan'. In this tale, the Devil is depicted as 185cm (6ft 1in) and 400kg (900lb) of pure radium contained within a polonium skin. He is a 'statue of pallid light. . . an incandescent glory'. The mortal who has summoned Satan is enamoured of polonium's strange light, as were the Curies. But one person's enchanting glow is another's stark warning. The pale blue glow is the result of the air around it being ripped apart by the radiation it is giving off. It is why microscopic amounts of polonium are sometimes used in anti-static devices.

And Twain's Devil has more tricks up his sleeve. Satan can light a cigar with a touch of his fingertip. The immense energy generated by polonium's radioactive decay also comes in the form of heat. A capsule containing only one gram of polonium will reach a temperature of 500°C (930°F). It means the element can be used to keep moon rovers warm during the cold lunar nights.

Mark Twain's 'softly glowing, richly smouldering' Satan is only a hint of the potent power within. The Devil's threat to humankind is embodied in his very being: 'If I should strip off my skin the world would vanish away in a flash of flame and a puff of smoke'. Twain got many scientific details wrong, but he was remarkably prescient about the power of radioactive elements.

Once the excitement over polonium's discovery subsided, with few applications to its credit and considerable downsides counting against it, the element was largely forgotten. It was too rare to be a concern to public health and, for most people, polonium became little more than another square on the periodic table. But not everyone had forgotten about it. In 2006, polonium created another political storm, when it was used to poison a former KGB agent living in the United Kingdom.

If polonium can rip apart the very air and keep rovers toasty warm in deep space, it is no stretch to appreciate the damage it can do inside a human body. The maximum safe body burden is an unimaginably tiny seven picograms (0.0000000000007 grams), making polonium a trillion times more lethal than hydrogen cyanide. Alexander Litvinenko swallowed an estimated 100 mg (0.1 grams) of polonium chloride, a poison his killers had perhaps hoped would be too obscure to trace. But polonium's intense radioactive decay left a trail of contamination that led to Russia and a marked political difference of opinion over what had happened, who was responsible and what should be done about it. Public awareness of this element had not been so high for over a century, but people's opinion of it had never been so low.

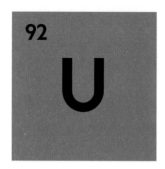

92

U

Uranium
The Game Changer

The periodic table can be read like a book. Starting at hydrogen and moving from left to right along the rows, like lines on a page, new characters emerge, and trends and patterns develop. There are a few surprises along the way, but the story of the elements progresses in a very predictable fashion. Then, just as you settle into a regular rhythm, you arrive at uranium. The ninety-second instalment of the elements saga is the plot twist at the end of the periodic tale.

From a historical perspective, uranium's story was set up perfectly. It started off as an extra, a background feature that made the scene more interesting to look at. Without knowing exactly what it was, the ancient Romans were using uranium oxide ores to make yellow glazes for their ceramics. In the Middle Ages, craftsman made yellow glass from pitchblende mined in Bohemia.

In 1789, uranium was promoted. It became a named but minor character in a large cast. German chemist Martin Heinrich Klaproth, believing he had isolated a new metal from pitchblende, called it uranium after the planet Uranus. In fact, he had isolated uranium oxide, and pure uranium metal was first produced by French chemist Eugène-Melchior Péligot in 1841. The metal was noted for its strength

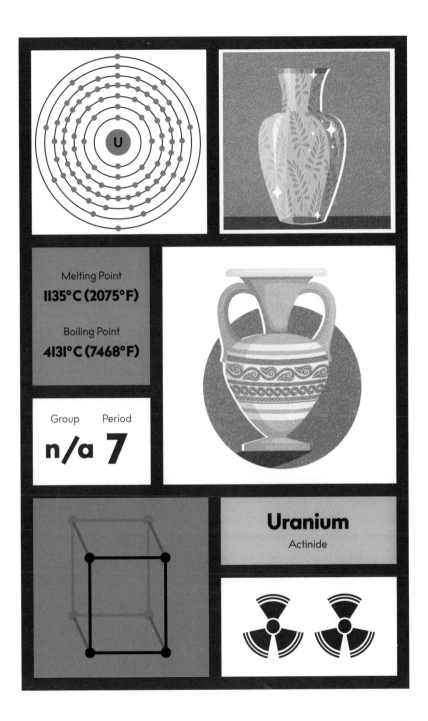

Melting Point
1135°C (2075°F)

Boiling Point
4131°C (7468°F)

Group
n/a

Period
7

Uranium
Actinide

but remained overshadowed by other more glamorous or useful elements. When Dmitri Mendeleev came to arrange all the known elements in his famous table in 1869, uranium was a familiar, but uninspiring member. It was most notable for being the end of the elemental line, the full stop at the end of the periodic table. Nothing was expected to exist beyond it. But then everything changed.

In 1896, French physicist Henri Becquerel saw another side to uranium. A sample of uranium salts left on top of a stack of photographic plates caused them to fog, even though they had not been exposed to light. The uranium was emitting something, some 'invisible rays' that could travel through the covering over the plates to expose them. Becquerel had discovered radioactivity. The story of the elements had not ended. Instead, it was about to take a sharp detour.

Over the decades that followed Becquerel's discovery, scientists picked apart the components of an atom: the protons that defined the element, the neutrons that held the protons in place, and the electrons that spun around them balancing everything out. Heavy atoms, ones with a lot of protons or an unstable number of neutrons, could eject unwanted extras along with a lot of energy. This was the radiation Becquerel had observed. And, some wondered, if atoms could eject particles, perhaps they could also accept them.

In 1938, Italian physicist Enrico Fermi attempted to push the limits of the periodic table by firing atomic particles at uranium atoms to see if they would stick. He claimed discovery of element 94, and was awarded a Nobel Prize for it, but he had got it wrong. Fermi had discovered something far more important. His uranium atoms had not absorbed the particles, they had been smashed apart, releasing phenomenal amounts of energy. It was the key to harnessing atomic power.

Element 94 was duly discovered in 1940, by a team at Berkeley University, again using uranium as the starting point, but with a modified method. The new element, named plutonium, was of particular interest at the time because it was the perfect material for making an atomic bomb. The problem was making enough of it to create a viable weapon. Uranium was again the answer.

In theory, packing enough uranium together in a small space would set up a chain reaction. When one uranium atom decayed, the particle it ejected could be taken up by another uranium atom causing that to decay and spit out more particles, and so on. Uranium, theoretically, could breed plutonium. All the team needed was the space and the uranium to try it.

Uranium sources and scientific resources were funnelled into research that was simultaneously cutting edge and makeshift. The first nuclear reactor was constructed in a squash court located under a football stand in the heart of Chicago city. The world has never been the same since. The atomic age has brought with it the threat of nuclear war and the promise of cheap energy from nuclear power stations. Uranium is now hot property. It has changed the way we see the periodic table and the world, a plot twist no one saw coming.

Plutonium
The Hell-Raiser

Plutonium is the angry young man of the periodic table. It seems resentful of ever being created. Plutonium is not just upset, its different forms run the full spectrum of anger from seething resentment to explosive violence.

When the team at Berkeley, led by Glenn Seaborg, decided to look for new elements beyond the supposed end of the periodic table, they were heading into uncharted territory. They knew that any elements heavier than uranium, if they existed, were likely to be unstable, but they could not have predicted just how uncooperative and temperamental element 94 would turn out to be.

An element's identity is determined by the number of protons in its nucleus. But the number of protons in an atom is not necessarily fixed. Unstable atoms can reconfigure the particles in their nucleus and eject unwanted ones to make themselves more comfortable. And a scientist with the right piece of kit can fire particles at atoms to alter the contents of their nuclei artificially. By firing particles at atoms of element 92, uranium, Seaborg and his team produced element 94.

Discovering an element means you also get to suggest a name for it. Seaborg chose plutonium, after the tiny planet discovered only eleven

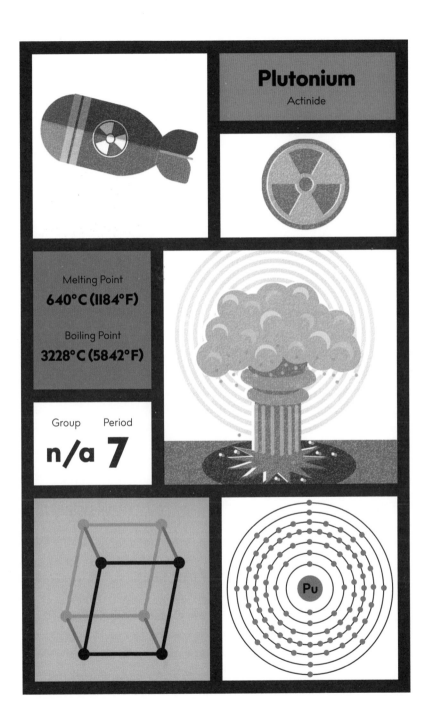

Plutonium
Actinide

Melting Point
640°C (1184°F)

Boiling Point
3228°C (5842°F)

Group
n/a

Period
7

Pu

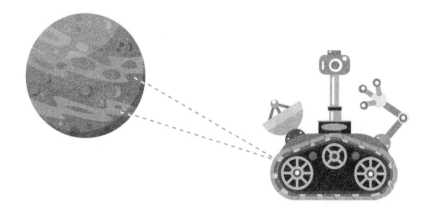

years earlier, orbiting at the extremes of our solar system. It seemed fitting. But Pluto had been named after the Roman god of the underworld, something Seaborg claimed not to know at the time. It turned out to be far more appropriate than he realized. Plutonium is every bit as infernal as its namesake.

Seaborg described plutonium as 'so unusual as to approach the unbelievable'. Even when he went on to describe some of the element's extraordinary properties, he was still underselling it: 'Under some conditions it can be nearly as hard and brittle as glass; under others, as soft and plastic as lead. It will burn and crumble quickly to powder when heated in air, or slowly disintegrate when kept at room temperature. . . It is unique among all the chemical elements. And it is fiendishly toxic, even in small amounts.' And Seaborg was only describing plutonium in general, things get worse when you start to examine the different varieties of this element.

The 94 protons inside every plutonium nucleus each have a positive charge. Positive charges repel each other and so neutrons, neutral particles as their name suggests, are also needed to hold them all together. The number of neutrons in a nucleus makes no difference to an element's identity. Plutonium, whether it has 144 or 150 neutrons, is still plutonium, they are just different varieties, known as isotopes.

Plutonium-238, with 144 neutrons, is a ball of hatred raging against everything, but to relatively little effect. This isotope is a powerful emitter of radiation, meaning blocks of it glow red hot. However, the radiation it emits does not get very far and can be stopped by a single sheet of paper. Plutonium-238's pent up energy has been channelled into fuelling satellites and space vehicles that explore our solar system.

By contrast, plutonium-240 is too unstable even for a nuclear weapon. Plutonium-240 will not just spit out unwanted particles, it will fall apart at the least provocation. Its untimely disintegration will cause a bomb to fizzle out rather than detonate.

Between these two extremes is plutonium-239. It has a fearsome temper, but it takes the right kind of goading to get it going. Like its 240 counterpart, plutonium-239 can be broken into fragments, releasing huge amounts of energy, but only with considerable provocation from being bombarded by neutrons. Spread out your plutonium-239 atoms and control the number of neutrons flying around and you have a reliable energy source that can be used to generate electricity. Pack the same plutonium-239 atoms tightly together and you have a bomb.

On 16 July 1945 a sphere of plutonium-239 was squeezed together to detonate the world's first nuclear explosion. The scientists involved in its design took bets on how big the resulting explosion would be. The physicist J. Robert Oppenheimer predicted it would be the equivalent of 300 tonnes of TNT. In fact, just over 6kg (13lb) of plutonium-239 exploded with the force of 15,000 tonnes of TNT. As Oppenheimer watched the mushroom cloud reach 12km (7.5 miles) into the sky, he recalled a quote from the ancient Hindu scripture the Bhagavad Gita: 'Now I am become death, destroyer of worlds.'

Mendelevium
The Chemist

The periodic table is the simplest illustration of the fundamental principles of chemistry. It is printed in every chemistry book and adorns the wall of every chemistry lab because it is the first lesson, the 101 guide to chemistry. Each element has its place on the table represented by a one or two-letter symbol in a box, often with a few other details included, such as mass, atomic number and full name. The 101st element included on every one of those tables is perhaps the most appropriately named element of them all, for it is named after the man who devised the periodic table, Dmitri Mendeleev. His insights into chemistry laid the foundations of the science that enabled his elemental namesake to be created.

Elements had been identified and defined long before Mendeleev was born. Lists of these elements had been compiled and they were often ordered by increasing mass. Some scientists had even noticed similarities and coincidences between certain elements and tried to arrange them together in small groups. But Mendeleev was the first to see the big picture. He realized there was more to it than a few odd coincidences between a handful of elements. There was an underlying regular, or periodic, system to the chemical behaviour of all the elements.

Mendelevium

Actinide

Melting Point
827°C (1521°F)

Boiling Point
Unknown

Group   Period
n/a   7

Md

The periodic table, arranged as it is at the front of this book, owes everything to Mendeleev's brilliant elemental insights. That he achieved this without knowing anything about what an atom was or how it was constructed, and with a large number of the elements as yet undiscovered, makes his feat all the more impressive.

His starting point, like that of many others before him, was mass. He placed the lightest element, hydrogen, at the top of the list, and followed on with the next lightest, and the next. As the list grew, he noticed recurring themes. Seven elements after lithium was another element that was very similar, sodium. Another seven elements further on was potassium that also shared many of the same chemical and physical properties with lithium and sodium.

Rather than continue with one long vertical list, Mendeleev made a table eight columns wide and started to add the elements in mass order from left to right. Every time he came across an element that was similar to the first row, he added it underneath, so each column was filled with similarly behaving elements.

Problems started to become apparent with this scheme. Sticking rigidly to the rule of increasing mass, some elements ended up in rows where they clearly did not belong. Tellurium is heavier than iodine, but their chemical characteristics suggest it should be the other way round. Mendeleev asserted it was his periodic system that was right and everyone else must be wrong. He changed tellurium's mass to what he thought it should be and confidently added it to his growing table to the left of iodine.

Another problem was that there were clearly bits missing from the table – like the answer in a crossword that does not quite have enough letters. At this point, many would think they had got the wrong answer to the clue. If Mendeleev ever questioned whether he had got the right answer to the chemical clues before him, his doubts did not last. He simply spread out his elements and left some squares empty.

From Mendeleev's point of view, holes in his table did not correspond to holes in his theory. In fact, they bolstered his argument. These holes were gaps where elements should be, it was just that no one had discovered them yet. Mendeleev could use his brilliant insights into the nature of chemical characteristics to predict these elements' masses and properties before they had even been found. When they were finally tracked down, they displayed all the characteristics he had expected.

Mendeleev's periodic system was vindicated, and each new discovery only confirmed the correctness of his table. His simple method of organization clearly showed how elements had patterns of behaviour that could be used to predict the properties of new elements and the outcomes of chemical reactions. He showed that atomic weight determined the character of an element. The long sought-after goal of the alchemists might be achievable if only mass could be added or taken away from an element. In 1955 a team of scientists added together the masses of a helium and an einsteinium atom to make mendelevium and it had all the properties its namesake's periodic system predicted.

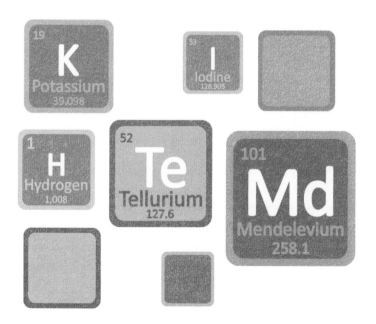

Meitnerium
The Tribute Maker

Names are important. They are not just there to label things so others can identify them, they can carry context, history, honour and shame. Choosing the right name for something as significant as a new element is not to be taken lightly. The name and symbol will appear on periodic tables in thousands of classrooms, laboratories and offices. It will bestow fame and recognition on the chosen person or place. You have to get it right.

In the late twentieth century a battle over the naming of a raft of elements, created in laboratories in the USA and USSR, grew so bitter it became known as the 'transfermium wars'. Element 100 had been named fermium after Enrico Fermi, the scientist who had fired the starting gun in the race to make new elements. Next, element 101, was equally uncontroversial as it was named after Dmitri Mendeleev, the father of the periodic table. Then things got tricky.

Whoever discovers a new element usually has the right to name it, or at least suggest a name that they hope will become widely accepted. But the new transfermium elements were being made only a few atoms at a time, and those atoms rapidly decayed into other elements. Proving they had ever existed was difficult. And, with high levels of mistrust on

Meitnerium

Transition Metal

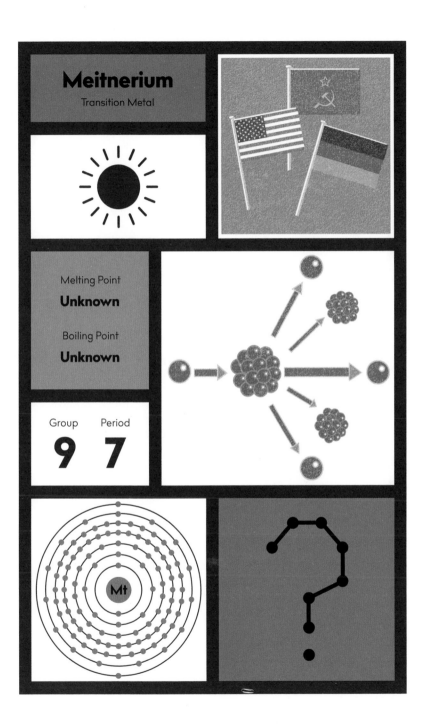

Melting Point
Unknown

Boiling Point
Unknown

Group **9**   Period **7**

Mt

both sides, the announcement of any new discovery on one side would be met with heavy scepticism from the other. Everyone became so entrenched in their views that, for a while, there were two versions of the periodic table. The Russian table had the Russian choices of names for elements 102 to 105; the Americans had their version, with their suggested names. One name, lawrencium, appeared on both tables, but for different elements. And all the time, more elements were being made. Germany entered the fray when they announced the discovery of elements 107 to 109 along with their own name suggestions. It was a confusing, embarrassing mess.

An independent committee was appointed to sort it all out, and in 1994 they revealed their proposed names for elements 102 to 109. For the first time in decades, all sides found something to agree on: they hated the committee's choices. New committees were appointed, and further proposals made. But while this ugly and occasionally farcical dispute rumbled on, one figure managed to stay above it all.

Lise Meitner was a brilliant scientist who had been forced to flee Nazi Germany because she was Jewish. Her colleague, Otto Hahn, was not Jewish, but was not a Nazi either. He gave Meitner his mother's diamond ring in case she needed to bribe her way out of Germany and continued to collaborate with her long-distance. When Hahn's experiments in Berlin produced results he could not explain, he wrote to Meitner in Sweden for advice.

Hahn had been trying to make new elements by firing atomic particles at atoms of uranium, element 92. Most of the particles would miss their tiny targets. Some might chip a few bits off a uranium nucleus, but a few should stick to make a heavier element. Though he tried everything, all he could find was barium, element 56. Hahn could not work out where the barium had come from or explain how a tiny particle could smash a barium-sized lump out of a uranium nucleus. So, he wrote to Meitner: 'Perhaps you can suggest some fantastic explanation.'

Meitner realized that if the nucleus of an atom was like a drop of liquid, firing a particle at it would make it would wobble and distort, and, in some circumstances, split into two smaller drops. In a series of long-distance phone calls, Meitner proposed Hahn conduct a simple experiment to confirm the theory. Meitner was right. She had correctly interpreted, not just Hahn's experimental results, but anomalous results other eminent physicists had completely misunderstood. The process, named fission, was a monumental discovery. It opened physicists' eyes to the potential of atomic power.

In 1945, a Nobel Prize was awarded for the discovery of nuclear fission. It went to Otto Hahn. Despite forty-eight nominations, Meitner never received the award. Many other honours have been given to her, but perhaps the most significant was in 1997. That year, the names of elements 102 to 109 were finally agreed. They now adorn every modern periodic table in the world. Element 109 is meitnerium, the only name that went undisputed in all the years of controversy. It marked the end of the transfermium wars and a fitting tribute to Lise Meitner.

118

Og

Oganesson
The Superheavyweight

Plenty of elements are a little unusual. Compared to their siblings, they may have a few quirks or act a little weird at times, but we usually simply accept their oddness because they are familiar. The further you move away from the everyday, the stranger things become. Elements at the far reaches of the periodic table are certainly strange, but their bizarre behaviour may only be a gentle introduction to the weirdness yet to come.

The furthest point of the periodic table, for the moment at least, is oganesson, element 118. It is named after Yuri Oganessian the man who has been instrumental in the discovery of the so-called superheavy elements that lie beyond element 103. Oganesson may be like Alice poised on the mantelpiece, about to step through the looking glass.

Regions beyond uranium in the periodic table have been likened to a map, where islands represent stable elements, surrounded by a sea of instability. Some of these islands are big, some are little more than sandbars barely poking above the waves. Element hunters, like adventurers, hope to wash up on these shores to hunt down exotic species.

For a long time, element hunters were able to make good predictions about what they were likely to find in their explorations. The arrange-

Melting Point
Unknown

Boiling Point
Unknown

Group   Period
18   **7**

Oganesson
Noble Gas?

ment of the periodic table illustrates trends and commonalities between elements. It all comes down to an atom's electrons. The electrons of an atom are arranged into shells and sub-shells. There are strict rules as to the shape and organization of these shells and sub-shells that determine how one atom will behave when it meets another one. Elements with the same pattern of electrons in their outermost shell will behave in similar ways. For example, tin has much in common with lead because both elements have a similar arrangement of outermost electrons.

When flerovium, element 114, was discovered, the obvious place to put it was next to element 113 and just below lead. Being below lead meant it should behave a little like lead, but flerovium, named after Georgy Flyorov, another giant in the world of element hunting, does not seem to be playing by the rules. The problem is the electrons. These superheavy elements have so many electrons that their shells have swelled to an enormous size to accommodate them, causing all sorts of problems. Rather than confining themselves to well-ordered configurations, they swarm about like an unruly mob.

If oganesson stuck to the rules, behaving like the other members of its periodic family, it would be an unreactive gas. Many scientists expect it to be both a solid and reactive. But, when only a handful of oganesson atoms have ever been produced, and they decay within fractions of a second, educated guesses about its chemical behaviour are likely to remain just that, at least for the time being.

While some scientists try to tease out flerovium and oganesson's chemical secrets, and those of the elements in between, the race to find more elements has not stopped. The next targets are elements 119 and 120 and they are expected to be tracked down in the not-too-distant future. Where scientists will place them in the table is less predictable.

Oganesson may mark a line in the periodic sand, or Alice teetering at the entrance to a rabbit hole. Beyond element 118 all bets are off. Will 119 and 120 settle into the spaces below groups 1 and 2 on the far left? Or will the two rows of elements underneath the main table sprout a third? Maybe these new elements will be so strange they will need to be kept separate from the rest of the 'conventional' elements. Perhaps, in the coming years, the whole periodic family tree will need to be redrawn. No one knows.

Element 120 is unlikely to be the end of the adventure. There are expected to be many more islands of stability out there to explore. Some scientists think the periodic family could extend to 172 members and they are likely to include elements every bit as strange as Lewis Carroll's bandersnatch, jubjub or jabberwock. Like Carroll's crew of ten who set off in search of a snark, a creature that could have feathers and bite, or have whiskers and scratch, element hunters are not quite sure what they are looking for. What is certain is that these elements, just like the snark, will not be captured in a commonplace way.

Timeline of Discovery

● ● ● ● = elements included in the book

● pre-history	6 Carbon	1790	38 Strontium
● pre-history	16 Sulphur	● 1791	22 Titanium
● pre-history	29 Copper	● 1794	39 Yttrium
● ancient	82 Lead	● 1797	4 Beryllium
● c. 3000 BCE	47 Silver	● 1798	24 Chromium
● c. 3000 BCE	79 Gold	1801	23 Vanadium
● c. 2500 BCE	26 Iron	1801	41 Niobium
● c. 2100 BCE	50 Tin	1802	73 Tantalum
● c. 1600 BCE	51 Antimony	1803	45 Rhodium
● c. 1500 BCE	80 Mercury	1803	46 Palladium
● pre 20 BCE	30 Zinc	1803	58 Cerium
● 1250	33 Arsenic	1803	76 Osmium
● c.1500	83 Bismuth	1803	77 Iridium
● 1669	15 Phosphorus	● 1807	11 Sodium
● pre 1700	78 Platinum	● 1807	19 Potassium
● 1735	27 Cobalt	● 1808	5 Boron
● 1751	28 Nickel	● 1808	20 Calcium
● 1755	12 Magnesium	● 1808	56 Barium
○ 1766	1 Hydrogen	● 1811	53 Iodine
● 1772	7 Nitrogen	● 1817	3 Lithium
● 1774	8 Oxygen	● 1817	34 Selenium
● 1774	17 Chlorine	1817	48 Cadmium
1774	25 Manganese	● 1824	14 Silicon
1781	42 Molybdenum	● 1825	13 Aluminium
1783	52 Tellurium	● 1826	35 Bromine
● 1783	74 Tungsten	1829	90 Thorium
1789	40 Zirconium	1839	57 Lanthanum
● 1789	92 Uranium	1843	65 Terbium

1843	68 Erbium	1925	75 Rhenium
1844	44 Ruthenium	● 1937	43 Technetium
1860	55 Caesium	1939	87 Francium
● 1861	37 Rubidium	1940	85 Astatine
1861	81 Thallium	1940	93 Neptunium
1863	49 Indium	● 1940	94 Plutonium
● 1875	31 Gallium	1944	95 Americium
1878	67 Holmium	1944	96 Curium
1878	70 Ytterbium	1945	61 Promethium
1879	21 Scandium	1949	97 Berkelium
1879	62 Samarium	1950	98 Californium
1879	69 Thulium	1952	99 Einsteinium
1880	64 Gadolinium	1953	100 Fermium
1885	59 Praseodymium	1955	101 Mendelevium
1885	60 Neodymium	1963	102 Nobelium
● 1886	9 Fluorine	1964	104 Rutherfordium
1886	32 Germanium	1965	103 Lawrencium
1886	66 Dysprosium	1968–70	105 Dubnium
1894	18 Argon	1974	106 Seaborgium
● 1895	2 Helium	1981	107 Bohrium
● 1898	10 Neon	● 1982	109 Meitnerium
1898	36 Krypton	1984	108 Hassium
1898	54 Xenon	1994	110 Darmstadtium
● 1898	84 Polonium	1994	111 Roentgenium
1898	88 Radium	1996	112 Copernicium
1899	89 Actinium	1999	114 Flerovium
1900	86 Radon	2000	116 Livermorium
● 1901	63 Europium	2004	113 Nihonium
1907	71 Lutetium	● 2006	118 Oganesson
1913	91 Protactinium	2010	115 Moscovium
1923	72 Hafnium	2010	117 Tennessine

Bibliography

Aldersey-Williams, H. 2012. *Periodic Tales: The Curious Lives of the Elements.* Penguin Books, London.

Bryson, B. 2004. *A Short History of Nearly Everything.* Transworld Publishing, London.

Burnett III, Z. 2020. LSD Perfume and Exploding Seashells: The 40-Year History of CIA Plots to Kill Castro. *MEL Magazine,* https://melmagazine.com/en-us/story/cia-plots-kill-fidel-castro

Carroll, L. 1876. *The Hunting of the Snark: An Agony in Eight Fits.* Macmillan and Company, London.

Chapman, K. 2019. *Superheavy: Making and Breaking the Periodic Table.* Bloomsbury Publishing, London.

Chaston, J. C. 1980. The Powder Metallurgy of Platinum: An Historical Account of its Origins and Growth. *Platinum Metals Review,* volume 24, issue 2.

Davy, H. 1806. The Bakerian Lecture, on Some Chemical Agencies of Electricity. *Philosophical Transactions.*

Emsley, J. 1989. *The Elements.* Clarendon Press, Oxford.

Emsley, J. 2001. *Nature's Building Blocks: An A-Z Guide to the Elements.* Oxford University Press, Oxford.

Emsley, J. 2001. *The Shocking History of Phosphorus.* Pan Books, London.

Everts, S. 2016. Van Gogh's Fading Colours Inspire Scientific Enquiry. *Chemical and Engineering News,* volume 95, issue 10.

Freestone, I., Meeks, N., Sax, M., Higgit, C. 2007. The Lycurgus Cup – A Roman Nanotechnology. *Gold Bulletin,* volume 40, issue 4.

Golomb, B. A. 1999. *Pyridostigmine Bromide, chapter 10: Bromism.* National Defense Research Institute, RAND.

Gusenius, E. M. 1967. Beginnings of Greatness in Swedish Chemistry: Georg Brandt (1696–1768), *Transactions of the Kansas Academy of Science,* 70(4): 413-425.

Hager, T. 2008. *The Alchemy of Air: A Jewish Genius, a Doomed Tycoon, and the Scientific Discovery That Fed the World but Fuelled the Rise of Hitler.* Three Rivers Press, New York.

Kauffman, G. B., Mayo, I. 1993. Memory Metal, *ChemMatters,* October, P4.

Kean, S. 2011. *The Disappearing Spoon: And Other True Tales from the Periodic Table.* Transworld Publishers, London.

Klaassen, C. D. (ed). 2013. *Casarett & Doull's Toxicology: The Basic Science of Poisons.* McGraw-Hill Education, New York, Chicago, San Francisco.

Lane, N. 2002. *Oxygen: The Molecule That Made the World*. Oxford University Press, Oxford.

Monico, L., Van der Snickt, G., Janssens, K., et. al. 2011. Degradation Process of Lead Chromate in Painting by Vincent van Gogh Studied by Means of Synchrotron X-ray Spectromicroscopy and Related Methods. 1. Artificially Aged Model Samples. *Analytical Chemistry*, volume 83: 1214-1223.

Principe, L. M. 2013. *The Secrets of Alchemy*. The University of Chicago Press, Chicago and London.

Rayman, M. P. 2000. The Importance of Selenium to Human Health. *The Lancet*, issue 356.

Relman, A. S. 1956. The Physiological Behaviour of Rubidium and Caesium in Relation to that of Potassium. *Yale Journal of Biological Medicine*, volume 29, issue 3.

Sacks, O. 2001 (2012). *Uncle Tungsten*. Macmillan Publishers Limited, London.

Scott, D. 2014. *Around the World in 18 Elements*. Royal Society of Chemistry, Cambridge.

St Clair Thomson, Sir. 1925. Antimonyall Cups: Pocula Emetica or Calices Vomitorii. *Proc Roy. Soc. Med.*, volume 19, issue 9.

Stone, T. and Darlington, G. 2000. *Pills, Potions and Poisons*. Oxford University Press, Oxford.

Twain, M. 1904. *Sold to Satan*.

Van Dyke, Y. 2015. *The Care of Lead White in Medieval Manuscripts. Toronto Art Restoration*. https://torontoartrestoration.com/the-care-of-medieval-manuscripts-issues-with-lead-white/

1967. *The First Weighing of Plutonium*. University of Chicago.

Websites
Royal Society of Chemistry Interactive Periodic Table:
https://www.rsc.org/periodic-table
http://drinkcrazywater.com/crazy-water-history/

About the Author

Kathryn Harkup is a chemist and author. Her first book was the international best-seller, *A is for Arsenic*. She has also written about the science of Frankenstein in *Making the Monster*, all the ways to die in a Shakespeare play in *Death By Shakespeare* and investigated the scientific background to horror's most famous fiend in her most recent book *Vampirology*.

Acknowledgements

First of all a huge thank you to Kerry Enzor for inviting me to be part of such a fun project. It was all her brilliant idea and I am only jealous I didn't think of it first. I've thoroughly enjoyed revisiting some old elemental friends and meeting new ones. Enormous appreciation also goes to Anna Southgate for her editing, creative input and project wrangling to get everything together. Thanks also to Luke Bird for the fantastic design work and Jo Parry for the fabulous illustrations.

Writing is not always such a solitary existence. There are many people who are invaluable to the process because of their constructive criticism, and their keen eye for bad grammar, poor phrasing or general nonsense. Paul Carpenter, David and Sharon Harkup, Dónal Mac Erlaine, Ashley Pearson and Richard and Violet Stutely have all kindly given their time and considerate responses on what I have written. As always, my parents, Margaret and Mick, have uncomplainingly (at least to me) read every word I have written, whether they are interested in the topic or not, and given helpful feedback. Matthew May and Mark Whiting have both been extremely generous with their time as well as their chemical and metallurgical expertise. Everyone's feedback and advice has hugely improved the book. A big thank you to all of them. Any remaining errors, awkward sentences or incomprehensible passages are all down to me.

Thanks must also go to many of the Blue Sea Writers group for their encouragement and support throughout the whole writing process. As always, Bill Backhouse has been a brick. He's always there with tea and sympathy, even if it is virtually. Thank you.